Integrated Design and Environmental Issues
in Concrete Technology

RELATED BOOKS FROM TAYLOR & FRANCIS

Cement-based Composites: Materials, Mechanical Properties and Performance
  A.M. Brandt
Concrete in Hot Environments
  I. Soroka
Concrete in Marine Environments
  P.K. Mehta
Concrete Mix Design, Quality Control and Specification
  K.W. Day
Concrete Technology: New Trends, Industrial Applications
  Edited by R. Gettu, A. Aguado and S.P. Shah
Construction Materials: Their Nature and Behaviour
  Edited by J.M. Illston
Demolition and Reuse of Concrete and Masonry
  Edited by E.K. Lauritzen
Disposal and Recycling of Organic and Polymeric Construction Materials
  Edited by Y. Ohama
Durability Design of Concrete Structures
  Edited by A. Sarja and E. Vesikari
Durability of Concrete in Cold Climates
  M. Pigeon and R. Pleau
Global Warming and the Built Environment
  R. Samuels and D. Prasad
Guidelines for Landscape and Visual Impact Assessment
  Institute of Environmental Assessment and Landscape Institute
High Performance Fiber Reinforced Cement Composites 2
  Edited by A.E. Naaman and H.W. Reinhardt
Mechanisms of Chemical Degradation of Cement-based Systems
  Edited by K.L. Scrivener and J.F. Young
Non-metallic (FRP) Reinforcement for Concrete Structures
  Edited by L. Taerwe
Performance Criteria for Concrete Durability
  Edited by J. Kropp and H.K. Hilsdorf
Size Effects in Concrete Structures
  Edited by H. Mihashi, H. Okamura and Z.P. Bažant
Structural Design of Polymer Composites: Eurocomp Design Code and Handbook
  Edited by J.L. Clarke

Integrated Design and Environmental Issues in Concrete Technology

Proceedings of the International Workshop
'Rational Design of Concrete Structures under Severe Conditions'

Hakodate, Japan
7–9 August 1995

EDITED BY

K. Sakai

Civil Engineering Research Institute,
Hokkaido Development Bureau, Japan

CRC Press
Taylor & Francis Group
Boca Raton London New York

CRC Press is an imprint of the
Taylor & Francis Group, an **informa** business
A TAYLOR & FRANCIS BOOK

CRC Press
Taylor & Francis Group
6000 Broken Sound Parkway NW, Suite 300
Boca Raton, FL 33487-2742

First issued in paperback 2019

© 1996 Hokkaido Development Association
CRC Press is an imprint of Taylor & Francis Group, an Informa business

No claim to original U.S. Government works

ISBN-13: 978-0-419-22180-7 (hbk)
ISBN-13: 978-0-367-86462-0 (pbk)

A Catalogue record for this book is available from the British Library

Publisher's Note: This book has been prepared from camera ready copy provided by the individual contributors.

**Visit the Taylor & Francis Web site at
http://www.taylorandfrancis.com**

**and the CRC Press Web site at
http://www.crcpress.com**

Contents

Preface

A workshop on "Rational Design of Concrete Structures" was held at Hakodate, Japan, during 7-9 August 1995. This book resulted from the papers presented in the workshop. The purpose of this workshop was to discuss concrete technologies towards the 21st century. There were two significant points in the discussions. One was integration of structural design and durability design as a rational method of design. The other was concrete technologies in relation to global environmental issues.

It is presumed that the next century will be the century of the environment. Concrete will not be an exception in this regard. It is obvious that we need to re-systemize concrete technologies, and to accomplish this, a new direction of research on concrete will have to be pursued. From this point of view, 27 papers were presented by distinguished researchers, which led to fruitful discussions. Based on the discussions, the following "Hakodate Declaration" was adopted:

1. We, concrete experts, shall place environmental consciousness at the center of concrete technology towards the 21st century.
2. We, concrete experts, shall change the framework of concrete technology by integrating structural design and durability design and by considering planned maintenance.

I hope that this workshop contributes to progress in concrete technology towards the 21st century.

The Civil Engineering Research Institute(CERI), Hokkaido Development Bureau, is conducting joint research with the Norwegian University of Science and Technology. It should be noted that this workshop was held based on the bilateral agreement between the Japanese and Norwegian governments. Full acknowledgement should be extended to the Science and Technology Agency in Japan and the Japan International Science & Technology Exchange Center for their financial support.

Finally, I would like to express my sincere gratitude to all participants for their significant contributions in this workshop. I also would like to thank my colleagues at CERI for their devoted support.

Koji Sakai
Civil Engineering Research Institute
Hokkaido Development Bureau

Sapporo
June 1996

1 CONCRETE TECHNOLOGY IN THE CENTURY OF THE ENVIRONMENT

K. SAKAI
Civil Engineering Research Institute, Hokkaido Development
Bureau, Sapporo, Japan

Abstract

As we move toward the 21st century, many industries have begun significant endeavors to preserve the environment. There is no doubt that concrete engineers also have to consider related problems and to clarify the role of concrete. In this paper, a direction for concrete technology in the century of the environment is comprehensively discussed. Firstly, it is emphasized that studies are needed to clarify the interface between durability and safety in the design of concrete structures. Furthermore, the importance of high–performance concrete in the construction of structures with long lives is described with an example developed by the author's research group. Finally, a concept of environmentally–friendly concrete is provided. The outline of the author's research group project on "eco–concrete" is shown.
Keywords: Concrete technology, durability design, eco–concrete, global environment, high–performance concrete, structural design.

1 Introduction

We have to realize that a battle for the survival of human life has already begun. We should not, therefore, hesitate to take prompt action. A significant challenge for all industries in the 21st century is the conservation of the environment. Concrete is not exceptional in this regard. This means that we are now at a stage in which we have to change the framework of concrete technology. For example, the significance of long–term durability in the efficient utilization of concrete structures has to be emphasized in concrete practice. However, do we know the true factors which govern the durability of concrete? Have we rationally considered the interface between structural design and durability design? In addition, have we paid attention to symbiosis

Integrated Design and Environmental Issues in Concrete Technology. Edited by K. Sakai. Published in 1996 by E & FN Spon, 2–6 Boundary Row, London, SE1 8HN, UK. ISBN 0 419 22180 8.

or co-existence with plants, animals, insects, and other organisms? What potentials are there in concrete in this regard?

The purpose of this paper is to comprehensively discuss the direction of concrete technology in the century of the environment.

2 Concrete technology in the century of the environment

The more humans want to live in greater comfort and convenience, the more industrial ativities will develop further and larger amounts of resources and energy will be consumed, thus increasing the amount of industrial waste. When the total consumption of resources and energy was small, there were no problems or there were problems only in limited regions. However, with population increases and standards of living based on the consumption of large amounts of resources and energy, various distortions (i.e. global environment problems) have arisen.

There is a simple line of thinking that holds that concrete has destroyed the natural environment. However, is such a conclusion logical? Since human beings invented concrete, it has been used in great quantities to create comfortable living environments as well as to construct the infrastructure necessary for the the social systems that support our lives. The construction of structures requires changing the natural environment, and if building a structure is regarded as the cause of such destruction, it can be said that concrete has served to destroy the natural environment. However, it is an undeniable fact that environmental destruction is being accelerated due to delays in the establishment of infrastructure.

If concrete is seen as the cause of damage to the natural environment, it means that concrete's use by human beings has lacked wisdom. In fact, because concrete is less costly and more durable than other materials, it may be that concrete has not always been used properly. In other words, concrete has brought a great number of positive contributions to mankind, while it is also undeniable that its convenience has led to its inappropriate use without adequate consideration.

Despite the value of nature, assurance of a comfortable living environment comes first. Even without looking at the aftermath of the Great Hanshin Earthquake, we cannot go back to the state in which living outdoors is the norm. If so, we should start over and reconsider the future utilization of concrete in order to re-establish concrete technology.

Difficult global situations involve various problems, some of which are related to concrete. Global environmental problems, the roots of which are said to lie in the increase in population, extend over a very wide range. As engineers who are engaged in the concrete industry, we must first clarify the role of concrete in global environmental preservation and, if concrete itself or its utilization is a possible cause of environmental burden, formulate measures to mitigate it.

Thus, it is extremely significant to place environmental consciousness at the center of concrete technology when we manufacture and utilize concrete. Namely, in concrete technology in the century of the environment, the following things have to be emphasized:

1. Effective use of resources
   - Reuse of waste
   - Use of low–quality materials
   - Recycling of demolished concrete
2. Control of CO_2
   - Improvement of cement manufacturing methods
   - Use of admixtures
3. Development of high–performance concrete and its utilization
4. Development of environmentally–friendly concrete
5. Interfaces between durability design and structural design
6. Energy–saving construction methods
7. Rational maintenance system

In the following, rational design of concrete structures, high–performance concrete, and environmentally–friendly concrete are described.

3 Rational design of concrete structures

The basis for designing concrete structures is to secure the capacities required during their service life. The design of concrete structures is generally divided into durability design and structural design. Figure 1 shows the ideal design system for concrete structures. Environmental conditions are important as prerequisites to design. The environments to be considered are a) the global environment, b) the environments to which structures are exposed and c) amenity environments (aesthetics). Another important prerequisite is maintenance. Design levels differ according to future maintenance plans. To establish rational design methods for concrete structures, it is essential to clarify interfaces among these factors as shown in Fig. 1 and deal with them systematically.

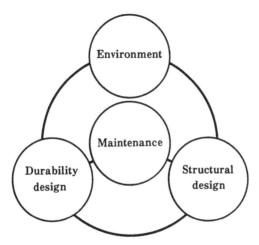

Fig. 1. Design system for concrete structures

Many studies on concrete durability have been conducted. However, it cannot be said that durability design which fully systematizes the results of these studies has been established. One of the reasons for this is that various factors affect the durability of concrete or concrete structures. In other words, it is believed that the following factors affect the durability of concrete or concrete structures.

1. Shapes of structures
2. Concrete cover and structural details
3. Properties of concrete (pore structure)
   - Mix proportions (water, water–cement ratio, water content, cement content and properties, admixtures, aggregates, etc.)
   - Workability
   - Surface treatment and curing
   - Workmanship
4. Environment

Deleterious substances which may be present in the environment include chlorides, carbon dioxide, oxygen, water, alkalis, sulfates and acids. These substances penetrate into concrete, chemically alter hydrates and corrode reinforcing bars. The types of deterioration include disintegration and/or dissolution of hydrated substances, micro- or macro–cracking, spalling and scaling. Besides these types of deterioration of concrete, cracking occurs due to plastic shrinkage or to strain caused by external forces, temperature changes, drying and other factors. The changes in mechanical properties by such deterioration affect the behavior of structural members, thus affecting the behavior of the overall structure.

Another obstacle to durability design is that no effective methods to evaluate durability have been established. It doesn't seem that such methods will be readily available in the near future. Although the permeability test is a possible method to evaluate durability, it is still in the development stage and has not yet reached the level of application to design. If durability of concrete can be properly evaluated and concrete of the quality required can be produced, then the mix proportion of concrete and details of concrete structures can be rationally determined according to environmental conditions. In other words, several design methods on different levels can be established, including one in which the deterioration of structures during their service life is not taken into account and one in which the deterioration which can be evaluated to some extent from an economic point of view is taken into account.

Thus, the realization of durability design depends on the establishment of effective methods to evaluate durability. It is greatly desirable to conduct studies with an emphasis on specific design methods.

Of the durability problems of concrete, the most important but least known is the quality of concrete in actual structures. The safety of concrete in structural design can be ensured by introducing a material factor (γ_c) to take into consideration variations in concrete quality. The compressive strength for design, which is used to check safety, can be represented by the following equation:

$$fcd = f'ck/\gamma_c \tag{1}$$

where f'ck represents characteristic compressive strength of concrete. This means that for safety, the required strength of concrete is f'ck but the expected strength of concrete is fcd. In other words, it is assumed that the actual concrete strength may be lower in a certain location than the characteristic strength of concrete. It can be believed that in fact such a phenomenon sometimes occurs, thus affecting the durability of concrete. Furthermore, the deterioration of concrete quality makes it easier for chloride ions to permeate concrete, thus affecting the safety of concrete structures due to the corrosion of reinforcing bars. It can be said that the handling of this problem is an "interface problem" between structural and durability design.

The "Standard Specification for Design and Construction of Concrete Structures"[1] (Japan Society of Civil Engineers) shows that a material factor for concrete ($\gamma_c = 1.3$) takes into consideration the extent of quality control, differences between the strengths of specimens and of material in a structure, changes in the concrete over time, and other factors. The changes in concrete over time are basically considered to be deterioration. Such changes greatly depend on the quality of concrete produced. Therefore, it is not reasonable to consider these things by using only one factor. Furthermore, it cannot be said that the factor ($\gamma_c = 1.3$) includes the effects such as those of reinforcing bar corrosion due to the permeation of chloride ions.

To incorporate the degree of changes of concrete or concrete structures over time into design the life–span factor can be introduced or a member factor (γ_b), which is used to examine the safety of the capacity of the cross section, can be used. Figure 2 shows the relationships among quality, the environment and life span for concrete structures. It is important to establish design methods by making models of the

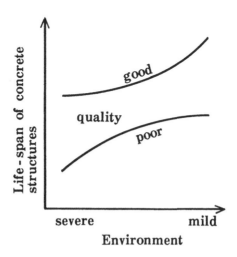

Fig. 2. Life–span of concrete structures

relationships based on data available now so that concrete structures can be constructed with an anticipated life span.

The design capacity of cross section (Rd) is represented by the following equation:

$$Rd = R(fd)/\gamma_b \qquad (2)$$

where γ_b is a factor which takes into consideration the degree of uncertainty of the calculated capacity of the member cross section, errors in the actual dimensions of members, the degree of importance of each member and fracture properties of members (e.g. γ_b is increased in the calculation of shear capacity for seismic design). The factor can also take into consideration the decrease of the capacity of members due to material deterioration. The factor for reducing the capacity of members, which depends on the quality of concrete, grades of construction and environmental conditions, can also be introduced into equation (2).

Even when using quality concrete with an emphasis on durability, no design systems have been established in which the advantages of quality concrete can be incorporated into structural design. In many standards, the maximum values of w/c are defined in terms of durability according to environmental conditions. When the upper limit values of w/c based on durability dominate, actual concrete strength may have a large safety margin as compared with the design strength of concrete. When the actual strength is greater than the design strength, the safety regarding shear capacity may decrease due to the increase in flexural capacity. Therefore, it is rational in such cases to conduct structural design by regarding the compressive strength obtained as the design strength of concrete. It can be said that such design is one that considers the interface between durability and safety.

Furthermore, even when w/c is defined, there is no guarantee that concrete with the quality expected actually will be obtained. Therefore, it may be necessary to introduce a safety factor on the maximum values of w/c themselves, depending on the importance of the concrete structure.

The deterioration of concrete results in changes in its mechanical properties. Therefore, to establish rational design methods which fulfill required durability and structural safety, it is necessary to obtain information on the relationship between micro and/or macro changes of the properties or material defects and the mechanical properties. Some examples are shown below.

Figure 3 shows the effect of cycles of freezing and thawing on the tensile properties of high–strength concrete [2]. Apparently freezing and thawing decreases the tensile strength. The tensile strength of concrete affects the occurrence of cracking, thus greatly affecting the durability of concrete. Figures 4(c) and (d) show the effects of bending as shown in Figures 4(a) and (b) on the resistance to freezing and thawing [3]. It can be seen from Figure 4(c) that micro–cracks due to bending scarcely affect the resistance to freezing and thawing because AE effectively works. However, it is shown in the case of non–AE that the larger the loading level becomes, the faster concrete deteriorates and that the degree of micro–cracks due to loading affects the rate of deterioration.

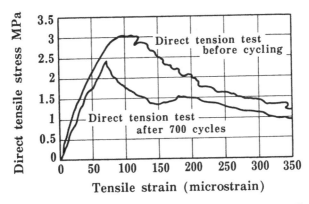

Fig.3. Effects of cycles of freezing and thawing on tensile properties of high–strength concrete [2]

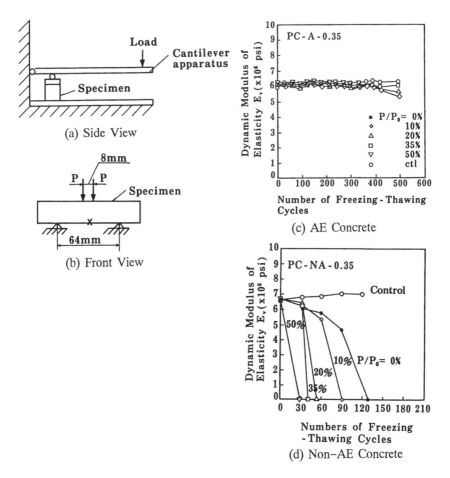

Fig. 4. Effects of bending on the resistance to freezing and thawing [3]

Figure 5 shows the effects of corroded reinforcing bars on the load–carrying capacity of reinforced concrete beams [4]. In the case of beams without lap splices, the load–carrying capacity decreases by about 10%. However, in the case of beams with lap splices, corrosion has an effect.

It is obvious that durability and safety have to be secured by integrating them in concrete structure design. However, researchers of concrete structures do not pay much attention to the effects of material deterioration. It seems that researchers of durability are not very interested in understanding how information on durability is utilized in the structural design. A small number of studies similar to the one above have been conducted. It is expected that such studies will grow more common. Without such studies, durability design and structural design cannot be integrated.

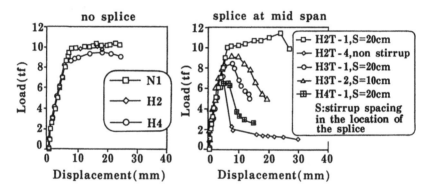

Fig. 5. Effects of corroded reinforcing bars on load–carrying capacity of RC beams [4]

The Great Hanshin Earthquake on January 1995, an earthquake of unprecedented scale in Japan's recent history, revealed that seismic design in Japan was not necessarily appropriate. It is expected that important structures will be designed to withstand earthquakes according to more stringent standards. However, it is also necessary to examine how the deterioration of materials affects the behavior of concrete structures which may be hit by earthquakes. Such examination is very important for the repairs and/or strengthening of the existing buildings.

4   High–performance concrete

Concrete engineering may be able to offer a solution for global environmental problems by constructing structures with long lives. There is much expectation of high–performance concrete (HPC) regarding the construction of such structures. High–performance concrete has a broad definition. Carino and Clifton [5] list the following characteristics as features necessary to define a concrete as "HPC":

1. Ease of placement and compaction without segregation

2. Enhanced long–term mechanical properties
3. High early–age strength
4. High toughness
5. Volume stability
6. Long life in severe environment (durability)

It can be said that high–performance concrete has no standard definition. The term was first used as a synonym for high–strength concrete. However, it is generally believed these days that high–strength concrete is not necessarily high–performance concrete because increasing the strength of concrete can cause other problems.

In general, high–strength concrete is produced by decreasing w/c with high–range water–reducing agents. However, this type of concrete has a higher proportion of cement, thus generating high temperatures during curing. Therefore, the author's study group has been involved in the development of a low–heat high–strength concrete [6,7]. Efforts were finally successful in developing an extremely high–performance concrete. This type of concrete is produced by mixing moderate–heat Portland cement or belite Portland cement, very finely–ground granulated blast–furnace slag, and silica fume. Figure 6 shows an example of the results. The compressive strength of 90 MPa was obtained at the age of 3 days. This type of concrete has the following characteristics: remarkable early–strength development, extraordinary strength development under low–temperature curing, and an extremely small adiabatic temperature rise, as shown in Figure 7.

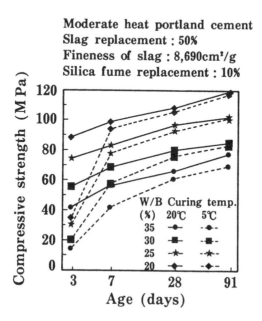

Fig. 6. Compressive strength of concrete developed by the Civil engineering Research Institute, Hokkaido Development Bureau

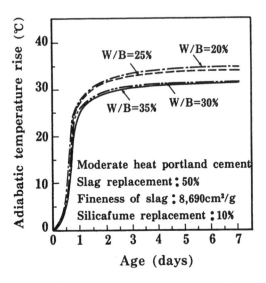

Fig. 7. Adiabatic temperature rise of concrete developed by the Civil
Engineering Research Institute, Hokkaido Development Bureau

The utilization of blast–furnace slag is an effective utilization of industrial waste. Reducing the consumption of Portland cement contributes to reducing the amount of CO_2 generated. Therefore, it can be said that this type of concrete is high–performance concrete which suits this century's emphasis on the global environment.

5   Environmentally–friendly concrete

With global environmental problems recently becoming serious, people have begun to move toward a new understanding of the importance of cycles of substances and of the food chain. Ecosystems exist in the balance between these factors. Therefore, people can understand the burden on the environment by returning to the starting point.

The recent development of materials is characterized by the idea that research and production are promoted not by seeking high–performance materials which consume large amounts of energy but by focusing on reducing the burden on the environment. "Life Cycle Assessment" (LCA), which aims to evaluate the burdens on the environment quantitatively and comprehensively at each stage of product production, consumption and disposal, has been gaining attention. Furthermore, the "Zero Emission Plan" of the United Nations University is being executed to transform the consumption–oriented industrial society that consumes large amounts of resources into a recycling–oriented one. The plan is a grand one which aims to create a waste–free social system.

Concrete is a construction material important to various structures. An important point in distinguishing concrete from ordinary materials is that many concrete structures come into direct contact with nature. This is one reason that, in many instances, people have a negative image of concrete structures. Concrete's hardness generally gives people the impression that the material is out of harmony with nature. In fact, few efforts to improve this situation have been made. To make concrete an eco–material, ways of understanding the situation and of solving the specific problems become important.

There is no doubt that locating concrete structures in certain areas results in changing the ecosystem in those areas, thus often making the lives of plants and animals more difficult. When symbiosis or coexistence of man and nature is considered, it is significant to evaluate the acceptable degree of environmental change. At the present time, it is very difficult to conduct such an evaluation. Therefore, the best method now is to give concrete a certain "interface function" at the point of contact between concrete and animals or plants to minimize the burden on the environment. Furthermore, it is desirable for concrete to have a form which can minimize interference with the cycles of substances, such as of water. When the limits of structural design prevent concrete from having such a form, it is important to compensate for the burdens.

When environmentally–friendly concrete is called "eco–concrete", eco–concrete can be defined as a type of concrete which contributes to reducing the burden on the global environment and which has ecologically–conscious interfaces between concrete and all living things, including humans. Figure 8 schematically shows the above concept. Porous concrete is such a type of concrete.

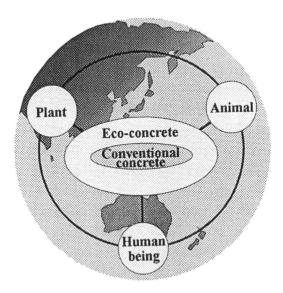

Fig. 8. A concept of eco–concrete

The author's research group has conducted an underwater exposure test of porous concrete specimens as shown in Photo 1 to clarify what role this type of concrete can play as a part of the underwater environment.

The dimension of specimens are φ 150 x 800 mm. Normal Portland cement and blast–furnace slag cement were used. Table 1 shows the mix proportions of concrete.

Photo. 1. Underwater exposure test specimens

Regardless of the types of cement and the sizes of aggregates, the percentage of voids was set at 25%.

Table 1 Mix proportions of porous concrete for underwater exposure

No.	Concrete	Type of cement	Size of coarse aggregate (mm)	Water–cement ratio (%)	Water (kg)	Coarse aggregate (kg)
1		Normal portland cement	20 ~ 40	28	88	1506
2	Porous concrete		2.5 ~ 5.0	28	103	1455
3		Blast–furnace slag cement	20 ~ 40	28	87	1506
4			2.5 ~ 5.0	28	101	1455
5	Normal concrete	Normal portland cement	25*	52	144	1193

* Gmax

Figure 9 shows the numbers of the aquatic insects that were commonly found in the specimens during 2 months of exposure. Many aquatic insects had already appeared during the first month of exposure. The number of aquatic insects clearly increased over time. It is evident that normal concrete, which has no voids, and porous concrete differ in the number of aquatic insects in the specimens. However, it is not clear whether the types of concrete and the size of voids affected the number of aquatic insects.

Thirty–six types of algae were observed during the first month of exposure. In particular, many diatoma elongatum and fragilaria construens were observed. The types of cement and concrete did not affect the number of algae. Observation during the second month of exposure basically showed the same results.

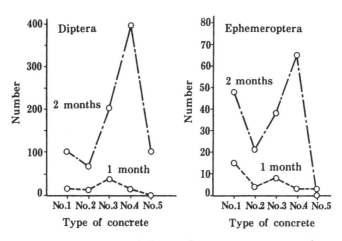

Fig. 9. Numbers of aquatic insects in porous concrete specimens

This project will be conducted for 2 years. This project incudes not only the observation of living things in the specimens but also measurement of water temperatures, pH, BOD(Biochemical Oxygen Demand), DO(Dissolved Oxygen), COD(Chemical Oxygen Demand), and SS(Suspended Solid). The effects will be evaluated comprehensively. There is no doubt that civil engineers must consider symbiosis with nature in carrying out their work. However, many obstacles hinder the realization of symbiqsis with nature. In this context, although the possibilities of porous concrete are as yet unknown, it is important to gather data steadily.

6   Concluding remarks

There is no doubt that the preservation of the global environment will be the most important problem of the 21st century. Therefore, concrete engineers must review concrete from this perspective. In other words, we will have to establish a new framework for concrete technologies. This paper focuses on a small number of problems. However, based on the discussions in this paper, subjects for future study include:

1. Establishment of a design system that integrates durability design and structural design
2. Development of high–performance concrete and clarification of the benefits of using this type of concrete

3. Development of methods to evaluate the durability of concrete
4. Development of methods to incorporate into design systems the life–span of concrete structures and the effects on the life–span of repairs and strengthening of structures
5. Clarifying the role of concrete in preserving the global environment
6. Development of eco–concrete

7   References

1. Japan Society of Civil Engineers. (1986) Standard Specification for Design and Construction of Concrete Structures, Part 1 [Design].
2. Marzouk, H. and Jiang, D. (1994) Effects of freezing and thawing on the tension properties of high–strength concrete. ACI Materials Journal, V. 91, No. 6, pp. 577–586.
3. Zhou, Y., Cohen, M.D. and Dolch, W.L. (1994) Effect of external loads on the frost–resistant properties of mortar with and without silica fume. ACI Materials Journal, V. 91, No. 6, pp. 595–601.
4. Katayama, S., Maruyama, K. and Kimura, T. (1995) Flexural behavior of RC beams with corrosion of steel bars. Extended Abstracts; The 49th Meeting of JCA, pp. 880–885.
5. Carino, N.J. and Clifton, R. (1991) High–performance concrete: research needs to enhance its use. Concrete International, V. 13, No. 9, pp. 70–76.
6. Sakai, K. and Watanabe, H. (1993) Development of low–heat high–strength concrete. Concrete 2000, (ed. R.K. Dhir and M.R. Jones), E & FN Spon, London, pp. 1539–1550.
7. Sakai, K. and Watanabe, H. (1994) High–performance concrete: low–heat and high–strength. ACI SP–149, pp. 243–268.

2 TOWARD RATIONAL DESIGN OF CONCRETE STRUCTURES – INTEGRATION OF STRUCTURAL DESIGN AND DURABILITY DESIGN

G.C. HOFF
Mobil Technology Company, Dallas, Texas, USA

Abstract
This paper discusses an approach that should be followed if the integration of durability design into the structural design process is to be successful. The inclusion of durability requirements in national codes and standards is probably the surest way to get uniform acceptance of these requirements by both designers and constructors. There is a definite need for shorter code and standard revision cycles, however. More definition of the environment in which the concrete structure will perform is necessary. Realistic service life requirements are necessary for every structure. Performance criteria should focus on the transport properties of the concrete as it is the ingress of the environment into the concrete which is the major problem. Combinations of both performance and prescriptive specifications are necessary to insure adequate durability. The evaluation of concrete performance should use national standards whenever possible rather than the vast array of tests proposed by many different researchers. Changes in the education process, both at the university level and for experienced engineers, would help remove resistance to changes in the design process. The need to include the durability considerations into the project planning is essential. Closer collaboration of the designers and the constructors with a view to eliminating structural details which can lead to durability problems is needed. The need for rigorous quality control and quality assurance on projects is essential to insure satisfactory durability in the future.
Keywords: Codes, concrete, constructability, detailing problems, durability criteria, education, environment, project timing, service life, specifications, standards, structural design, test methods.

Integrated Design and Environmental Issues in Concrete Technology. Edited by K. Sakai. Published in 1996 by E & FN Spon, 2–6 Boundary Row, London, SE1 8HN, UK. ISBN 0 419 22180 8.

1 Introduction

Concrete is the most extensively used material in the $400 billion per year construction industry in the U.S. It is also the most extensively used material in most other regions of the world. This common use contributes to the continuing problem of poor concrete durability as nobody wants to pay the extra attention and the little extra initial cost it takes to provide a more durable, longer lasting concrete. Because we use so much concrete, and because concrete has competing materials (eg., steel, wood, asphalt) for many applications, the tendency is to provide the concrete at the lowest possible cost. This is achieved by minimizing the materials required to produce the concrete, using the cheapest (and probably the least trained) labor, and eliminating the good practices (eg., quality control, consolidation, curing) that are necessary to produce good concrete.

What does this type of approach bring us? A 1990 National Research Council report in the U.S. indicated that it will cost $2 to $3 trillion over the next 20 years to repair all the U.S. concrete structures which are now deteriorating from corrosion or are poorly made and maintained. This is not to say that the competing construction materials survive any better. In fact, most do worse than even the poorest of concretes. A long running television commercial in the U.S. promoting the sale of automobile oil filters, concluded by having the auto repair mechanic saying "You can pay me now or you can pay me later!". The inference was that if you initially paid a little more for the oil filter that you used in your automobile, you wouldn't have to pay for costly repairs in the future. A similar analogy exists for concrete.

The need for a more durable concrete has not been lost on the global engineering community. As a result, most countries now have their own programs to develop "high-performance concrete" because they recognize that there is a great potential to minimize expense each year in maintenance, repair and replacement of poorer quality concrete. What exactly is high-performance concrete?

Many definitions have been proposed for "high performance concrete." In general, high performance means behavior above and beyond what we would normally expect from concrete. This tends to be somewhat of an anomaly because when we order concrete from a supplier, we expect it to have satisfactory performance with respect to both strength and durability. We certainly don't order "low performance" concrete but that is often what we get. We typically order concrete by strength requirements. Strength is quantifiable. If we order 40 MPa, we can measure it. How do we order durable concrete? If we ask for concrete that will last for 100 years, how do we or the concrete supplier know that it will? We won't be here to check and see that it has performed satisfactorily and, most likely, the concrete supplier won't be in business to compensate us for any non-performance if it fails in less than 100 years.

Indications are that high performance concrete is having a successful evolution, at least at the laboratory level. However, the money being spent on developing and evaluating high performance concretes is, for the most part, distributed among many researchers in many countries doing relatively small, uncoordinated studies. It is estimated that in the U.S. in 1994, there were 3000 concrete studies being done in 500 to 1000 laboratories or organizations. How this information gets disseminated and integrated into actual practice is a major problem.

Dissemination of information is a simpler problem than the integration of that information into the design process. Meetings of the type that this presentation is being made at, are helpful for information dissemination in that researchers are advising other researchers of their work and results. Normally researchers have trouble tracking what other researchers are doing as is evidenced by that fact that many papers on well researched durability problems (eg., reinforcing bar corrosion) reference only the work done on that subject by the author of that paper. However, the results from this meeting and the many similar meetings on concrete durability generally do not cause any significant changes in design or construction practice. Why is this? The following sections explore a few of the authors views on this problem.

It is also the author's belief that some of the major stumbling blocks to acceptance of durability considerations in design are education, timing of the introduction of durability considerations into a project, and the reluctance, by constructors, to change existing practices to accommodate durability requirements. All have a financial impact on a project as an insured durability comes at an initial higher cost. Some aspects of the influences of these topics are described below as catalysts for further discussion. Bryant Mather, world renowned expert on concrete durability, has often stated that based on what we know about concrete and its behavior in given environments, the only reasons concrete should deteriorate are ignorance, stupidity or fraud. Hopefully, we can change the first two of these within some reasonable period.

2 Implementation

Designers do not have a financial incentive to use the new technology being developed by the researchers on high performance concrete. Using or specifying new technologies tend to increase the designers cost and also the risk of doing business without any appreciable benefit or financial reward. A similar problem exists for the constructors as they generally are not a part of the design process and can only execute work in accordance with the documents of the contract. Occasionally, "value engineering" clauses exist in contracts which allow the constructor to use innovative construction materials and techniques to produce cost savings on a project. These savings are shared by the owner and the constructor, but these contract provisions are rare and there are numerous critics of this approach. For a constructor to deviate significantly from standard practice to implement innovations increases his risk which is a strong disincentive when most contracts are awarded on a least price basis.

When the designer is reluctant to specify new technology and the constructor is reluctant to implement it as his own discretion because of risk or cost, the typical tendency is to think of the government, at all levels, as the solution to the problem. Governments are in a position to lead in the development and demonstration of improved concrete technology by implementing constructions that use technology that may be outside of the provisions of codes and standards. This is due to the fact that governments are self- insuring and that they are entrusted with providing safe structures for public use using the publics money.

However, in many governments, the motivation for providing long-lasting materials does not exist. The principal motivators in taking action are a fear of disaster (eg., a bridge collapse) and lowest costs. You cannot sell the politicians on the concept of durability because the material only has to last as long as their term in office. Life-cycle costs are generally not considered. In private corporations, similar attitudes often exist because stockholders expect business to be performed at lowest costs. There are many examples of developers doing the minimum needed in construction because they do not intend to be the long term owner of the structure and hence do not care what happens to the structure in the years to come.

Would the designer, the constructor, and the government be more willing to adopt improved durability innovations if they could do it by reference to accepted codes, standards, and specifications? The answer is probably "Yes" because there is a general underlying faith by the industry and the general public in these types of documents. The codes and standards currently determine technology acceptance. If the accepted codes, standards, and specifications require it, then it puts all competing parties on a level playing field. Risks become the same and the durability innovations become part of the cost of doing business for everyone.

In support of this premise, the US Civil Engineering Research Foundation (CERF) conducted a survey of 600 members of the highway community, including public sector officials, construction companies, entrepreneurs, and research facilities to determine what were the most common barriers to adopting innovations in highway construction technologies. The ranking was 1) constrained standards or specifications (23%), 2) current limitations on proprietary product usage (18%), 3) lengthy processes or high costs (17%), and 4) known evaluation inadequate (12%). If the standards or specifications do not allow the use of new technology to improve the performance and service life of concrete, then that is a problem we certainly have the capability to fix.

3 The standardization process

Structural design of concrete typically never considers the durability aspects of the concrete. Strength is the primary consideration along with modulus of elasticity and other related mechanical properties. The integration of structural design and durability design can be approached from two directions: higher strengths lead to improved durability or improved durability leads to higher strengths. Neither approach is universally correct and exceptions can always be found. From a practical standpoint, it may be simpler to concentrate on improving the durability aspects as changes to codes and standards with respect to improved durability may be accomplished in a more timely manner than changes to structural provisions. Many structural requirements in codes and standards were developed for concretes of strengths 40 MPa or less. When higher strengths are used, these requirements may or may not become more conservative. Changes in the structural provisions need supporting data from tests of structural members and the production of such data is a long and expensive process, more so than for durability evaluations. However, if improved durability design results in routinely occurring higher strengths, the designers will, in time, adjust their procedures to effectively use this additional strength.

It is the author's belief that standards must be the place to begin to integrate durability requirements into the design process as many specifications simply refer to the standards. As noted above, constrained standards or specifications are a significant barrier to the implementation of new materials and technology into the design and construction process. Why are they constrained or restrictive? The existing design codes and standards are usually conservative because their underlying basis is to protect the public while providing a functional structure. They tend to restrict the use of new materials until, and unless, a substantial data base about them has been established. They are also flavored by the perception of what the industry can realistically produce. This is a dangerous perception because experience has shown that with a little training and education, high quality concrete production and construction can be achieved on a routine basis.

The preparation of codes, standards, and specifications are done differently in various parts of the world. Some countries use the "consensus standard" approach where representatives from all aspects of the industry, including academia, jointly prepare and approve the standards. Some countries use government agencies or teams of specialists to prepare the standards. Other countries simply adopt the standards of other countries and modify them for their local needs. All of these preparation methods run the risks of influence by special interest groups or individuals, or, even more commonly, having uninformed individuals either doing the standard preparation or approval or both. In general, these risks are reduced in the consensus standards approach because of the checks and balances provided by a very broad representation of the industry in the standards preparation.

How can we improve the standardization process to expedite durability requirements into the design process? The first step should be the timely and efficient evaluation of the durability data base being developed by the research and development community. This should be complimented by demonstration projects of the new technology when they exist. Unfortunately, when public funds are used, demonstration projects are not viewed favorably because they tend to benefit only a limited segment of the community instead of everyone who helped pay for it. The key word here is "timely". It is not very effective if the inclusion of new technology into standards takes 7 to 10 years, yet that time period is typical for many code and standard reissue cycles. A classic example is the preparation, in the US, of a standard for silica fume, a major contributing material for improving durability. After 10 years of development, Draft 16 of the standard is being balloted in 1995. Interim procedures for adopting new technology into standards must be invoked so that once a new technology or material has been satisfactorily demonstrated, it can appear in the standards within 12 to 18 months. This may be an impossible task in the short term but is something that needs serious consideration by the industry.

If we are to include more stringent durability requirements into our codes and standards, two approaches are possible: performance requirements and prescriptive requirements. Both have their advantages and disadvantages and perhaps combinations of the two can supply the best solution.

4 Performance and prescriptive specifications

Performance specifications are promoted as providing flexibility and encouraging economic mixture proportioning. "Economic mixture proportioning" infers that either the least amount of material will be used in producing concrete that will satisfy certain performance criteria or that less expensive materials than the conventional concrete materials will be used. The establishment of "performance criteria" assumes that we understand the nature of concrete and its interaction with both the environment and service loads. It also assumes that we can adequately define the environment and also predict the service life of the structure. This is doubtful although we are making progress in all of these areas. Realistically, we are only making educated guesses for most aspects of "performance criteria". There is also an uncertainty about the methods we use to measure "performance". While the concrete technologist may be aware of these limitations, the concrete designer assumes that if the concrete meets the performance criteria set forth in standards or specifications, the concrete when used in a structure in a given environment will have sufficient durability for the service life of the structure. Unfortunately, it is very difficult to provide a general criteria which meets all possible uses of concrete.

Prescriptive specifications, those that tell you exactly how much of each ingredient to put into the concrete, used to be common and still exist in many parts of the world. The mixing water is typically the only variable and is adjusted to suit the conditions of the construction. These would be ideal if all the ingredients used to make concrete were the same world-wide. They are not, so the best we can do, based on our present research on concrete durability, is to define minimum amounts of binder material and maximum amounts of water to be used with that binder so as to provide a dense matrix to resist the intrusion of the environment. We can also require portions of that binder to be supplementary cementing materials, e.g., fly ash, silica fume, and blast furnace slag. These supplementary cementing materials further densify the binder. The use of chemical admixtures that allow the reduction of water while maintaining adequate workability can also be prescribed. These admixtures may be necessary when using some of the supplementary cementing materials. Prescriptive specifications tend to be in conflict with the concept of "economic mixture proportioning" for performance specifications but are more easily understood by the user.

It is doubtful that more durable concrete can be achieved by minimizing the amounts of the more expensive ingredients of concrete (namely Portland cement) while still obtaining adequate strengths. As noted earlier, concrete strength is, however, the primary consideration for the designer. Strength is, at best, an indirect measure of concrete quality from a durability viewpoint. Historically, most standards and specifications have used the water-binder ratio as the main parameter to define concrete for durability, with the 28-day compressive strength consistent with the water-binder ratio providing the field control of the concrete. For marine concretes, minimum amounts of cementing material are also specified. The extensive research being done on high-performance concrete, however, indicates that all these requirements for improved durability, while necessary, may not be enough. What is needed is a very dense matrix of ingredients that are compatible with the aggregates and resistant to the environment in which the concrete is to used. In this regard, the use of low water-binder ratios, high binder contents, and supplementary cementing materials (eg., silica fume, fly ash, blast furnace slag) in conjunction with portland cement have typically shown superior performance to conventional concretes under many different environmental exposures. Most of these modifications to improve durability will not produce a more "economic" mixture. With the cost of the concrete materials being only 5 to 15% of the cost of the concrete in-place, it is also foolish to consider using insufficient amounts of material when considering what future maintenance costs will be.

5 The environment

There is one school of thought that believes that durability is not a property of the concrete but is a function of the environment. It postulates that all concretes are durable provided they are kept in an environment consistent with the composition of the concrete. Ancient concretes, made with natural cements, lasted for thousands of years in a benign environment. With the advance of civilization, the environment was changed as pollution increased and the concrete began to deteriorate. This can be witnessed in the ancient monuments in populated areas of Greece, Italy, Egypt and Mexico. Similar monuments in other regions of the world that have not seen significant pollution have not experienced this same level of deterioration. Concrete which survives in one environment may not survive in another environment. Concrete used in the interior of buildings may not survive when used in a marine environment, a sewage treatment plant, or for the containment of hazardous waste.

When concrete is put in a difficult environment, it will deteriorate at rates governed by that environment. If the rates of deterioration for that environment can be predicted, then sufficient material can be added to achieve a prescribed service life or the concrete constituents can be modified to reduce that rate and extend the service life. As an example, the corrosion of reinforcements has received extensive study for some time. Most of the principal factors that reduce the long-term integrity of reinforced concrete are well understood. The use of adequate concrete cover, dense concrete matrices, crack control measures, anti-corrosion admixtures and additives, and other techniques have resulted from this knowledge. Unfortunately, these are still relatively short term solutions. Few studies have been made that can help predict the longevity of reinforced concrete over hundreds of years. Very little information in the literature is helpful in relating, in a quantitative way, the rate of degradation of concretes subjected to the principal characteristics of the environment over the long term.

Do we know enough about our environments to establish a universal criteria for performance specifications? Our specifications presently address marine environments, sulfate soils, freezing and thawing, and a few other phenomena which can be quantified. But what about pollution, acid rain, hazardous wastes and other environments. While the present answer to a universal criteria is probably "no", I am confident that, with time, individual criteria can be established for almost all aggressive environments. If we had this information today, we would still be at risk in designing structures with very long service lives because we don't know what our environment of the future will be. If we design concrete to last 1000 years based on our environment today, do we have any guarantees that the environment will not become worse in the future? Hopefully it won't, but we can not be sure.

6 Service life

Service life of engineering structures has been a hot topic within the engineering community for the last several years, but it still lacks adequate definition. In general, service life of a structure is a questionable concept. Very few concrete structures have a precise service life. Offshore concrete platforms have a service life that is determined by the amount of hydrocarbon in the reservoir it is producing and the rate of production. Yet, concrete platforms built in the 1950's for a service life of 20 years are still in service because the technology of extracting the hydrocarbons from the earth has improved over the years so that much more is being produced than was originally planned for. Buildings designed for 40 years are still in service after 70 years while similar buildings are removed in 10 to 15 years to make way for more modern buildings or other facilities.

For the sake of the designer, do we need to specify a service life for a structure? The answer is "yes" even though we don't know how realistic it will be. Requiring that a structure should have a specific service life must also be accompanied by performance criteria which gives the designer a quantitative objective in the design process.

The service life of concrete is usually dependent on a slow rate of deterioration which results from combinations of the environment, the concrete quality, and the size and configuration of the concrete element. In addition, service life is also a function of the failure criteria which is used. A practical service life may also depend on the competition that concrete may receive from alternative materials. If a pavement is designed to last 100 years, it will probably never be built because the cost will not conform to most budgets of the agencies who fund highway construction. Cheaper materials than concrete will be used without regard to service life. Structures whose basic function and utility doesn't change with the passage of time, such as bridges, buildings, marine structures, dams and navigation structures, certainly should be designed to last at least 100 years. Other structures whose capacity will be exceeded within reasonable time periods because of increasing population or which will become obsolete due to changing technology, can have shorter lives. Whatever service life we specify, it should be on the conservative side because we can't anticipate all of the factors which might influence the structures longevity.

7 Performance criteria and measurements

Although properties such as compressive strength, modulus of elasticity, and porosity are indirectly related to durability, the transport properties of diffusivity and permeability are probably the most important factors relating to durability. Unfortunately, these properties are not simple things to measure. The major degradation parameters that exist in a service environment include sulfate ions, chloride ions, leaching of calcium hydroxide by water, carbon dioxide reactions, alkali-aggregate reactions, freezing and thawing, wetting and drying, and these various parameters in combination. If the transport of these degrading phenomena into the concrete can be reduced or eliminated, the long-term performance of the concrete can be greatly enhanced. Only by establishing criteria which demonstrates that the environment is not penetrating the concrete at unacceptable rates, can we make an estimate of the service life of the concrete.

Having once established the criteria, we must have realistic methods to evaluate the concrete to ensure conformance to the criteria. Traditionally, concrete durability has been assessed by measuring either the strength, weight change, length change or change in structure by non-destructive techniques of specimens subjected to a corrosive agent or a specific difficult environment. The corrosive agents are generally applied at the external boundary of the specimen but may sometimes be included in the concrete during mixing. Difficult environments typically include extremes of temperature and varying moisture conditions which are varied throughout the test, usually for time periods which are more suited to the test procedure than the real world environment. Results from these types of tests may not be applicable to predictions of long term durability and are generally not sensitive to the design geometry of structural members.

The durability renaissance of the 90's has led to an abundance of performance test methods for specifying durability as concrete technologists have endeavored to move away from strength as a durability guide. In a quick scan of test methods related to the permeability or diffusion rates of concrete, the author found 11 different methods by which these phenomena could be quantified. Tests of this type that have been developed in the laboratory using a small range of trial concretes may not translate well to other laboratories, other concrete, or the variety of environments that exist in the real world. Test specifications that have not been formulated by national associations may be flawed in technique or description. In our specifications, we should use national standard tests wherever possible.

Initially, including durability performance criteria and the methods by which to measure them in the specifications will meet a great resistance from both the designer and the constructor who are not used to working towards a "durable structure". The criteria will be "too severe". The test methods will be "impractical". These are legitimate concerns and caution must be exercised to ensure that what is asked for is absolutely necessary to obtain a durable concrete structure. The inclusion of impractical criteria and test methods can be an unacceptable hindrance to construction that leads to excessive costs and delays. Apart from impracticality, some "durability tests" may be no better an indication of performance than strength. The sudden implementation of arduous performance specifications can be expected to cause significant start-up problems on jobs and may prove impractical to enforce. These are not insurmountable problems but will require some education of the industry during the implementation period.

8 Education

In the years to come, the integration of durability considerations into the structural design process would proceed more smoothly if the design engineers had some basic concrete "know-how". Unfortunately, concrete "know-how" typically does not form a part of todays young engineers education and training. Neville [4] has observed that by "...not teaching something so ordinary, so simple as concrete (which could actually dirty your hands)...", the status of the teaching institution is actually enhanced. In many North American universities at the undergraduate level, the teaching of concrete fundamentals is typically combined in a "strength of materials" course along with many other types of materials. The total exposure to concrete is perhaps only several weeks. The student may subsequently be exposed to reinforced or prestressed concrete design studies, but design can be done with no real appreciation or understanding of the materials behavior. This is, of course, the underlying reason for this workshop. The real exposure to concrete appears to be at the graduate level. Unfortunately, the bright young engineers who go on to graduate school and specialize in concrete are very few with many of those remaining in academia rather than becoming designers.

It is often argued that there is not enough time in the curriculum to specialize in such a limited (and mundane) subject as concrete, yet there always appears to be time for other specialized subjects relating to our computerized (and perceived exciting) future. In a way, you can't blame the universities as they must offer a program that attracts students if they want to remain in business. On the other hand, concrete is the building material of choice in almost all regions of the world and certainly deserves more attention than it gets. Can we change the way universities establish their priorities and include a full 16 week course on concrete as the author had 38 years ago? Probably not in the short term, but we can at least begin trying.

Neville [4] also noted that the problem is not limited only to young engineers. More senior people may have basic "know-how" but typically are not in a position to mentor young engineers. They typically are in the office while the newer, younger engineers are in the field and have to learn basically by trial and error. Also, the more senior engineers generally have a certain unwillingness to accept newer (not new) methods and materials. The author dealt with a "senior" individual several years ago whose firm had just received a contract to produce and deliver over 165,000 m^3 of 80 MPa concrete.

He insisted that he could produce this concrete using cumulative weigh batchers and truck mixing as that is the way they always produced concrete in his area. Upon closer investigation, it was determined that his firm had never had never produced over 40 MPa concrete and his equipment and procedures were satisfactory for that quality of concrete. At the owners insistence, he converted to individual materials weigh hoppers, high shear stationary mixers, and improved quality control systems and now routinely produces 80 MPa concrete and wonders why he never did that before. Once he accepted a newer system to produce the concrete, he became "educated", in a small way, about the continuing improvements in concrete technology.

The problem of re-educating the more senior people is simpler than getting the basic education problem solved. Through the use of short courses (1 to 3 days) on specific subjects relating to the focus of this workshop, senior engineers can rapidly be informed of the new technology. These courses should be taught by concrete technologists and be organized and administered by National concrete organizations. The venue should be such that it is easily accessible at limited costs.

9 Timing

Consideration of the materials selection at the design stage must be given the same level of attention as given structural considerations. Common sense dictates that if a structural member must accommodate certain loads at given levels for the life of a structure, then the materials which comprise that member must maintain their integrity for the same time period. In many instances, however, the choices of ideal materials and techniques may clash with preconceived ideas on design and construction procedures employed by engineers. This goes back to the problem of education of the design engineers and also the owners. If the owner can be convinced that he must involve a concrete technologist at the earliest stage of his planning, then there should be sufficient time to get the materials evaluated so a proper selection can be made. Similarly, newer or innovative methods of construction can be considered and included in the bid documents for the work.

As an example from the authors experience, the same job noted above initially planned to use all local materials to produce the 80 MPa concrete. By involving a concrete technologist in the planning stage, it was determined that the local cement had an equivalent alkali content greater than 1.2% and that all local aggregates ranged from slightly to highly reactive to alkalis. The production of 80 MPa concrete also involved high cementing contents which put even more alkalis in contact with the aggregates. By initiating an alkali-aggregate reactivity evaluation program approximately 1-year before the award of contract, suitable materials for consideration by the contractor were identified. Upon award of the concrete, no delays in start-up were encountered because of materials selection. It was estimated that the cost of doing that preliminary work was only 0.03% of what the delay costs would have been.

10 Constructability

Many problems attributed to contractor deficiencies are caused by a lack of understanding of construction operations by the designer. These problems can be related to reinforcement details, section size and shape, embedments and penetrations, obstructions to casting or consolidation, and limited access for both construction and inspection. The designer may not be wholly at fault as the design may only reflect the Code requirements or the owners wishes. It is the authors opinion that a constructability review by people sensitive to the durability issues be made for every major construction. While reviews may be common practice in certain industries, the question of whether durability aspects are considered in those reviews is not always answered.

Simple details, such as having a slight slope on sills to avoid the collection of water or salt on them, can help avoid a potential problem of concrete deterioration. Yet, the designer will probably include a horizontal surface as it is easier to consider and the constructor will prefer to build a horizontal section as it is easier to construct. Many other simple changes for durability enhancement, such as rounded corners rather than orthogonal corners, could probably be identified if durability of the concrete was considered at the early stages of the project.

Poor construction practices will always persist if proper quality control and quality assurance procedures are not enforced. Training aids for concrete craftsman have been developed by several organizations and have been extremely beneficial in providing guidance to the workers on what is the proper way to do concrete construction. This training must be provided to the workers at the start of the project and reinforced as the project progresses.

With respect to concrete durability, the assumption that a qualified constructor knows what he is doing can get a structure in trouble very quickly. A constructor with a good reputation may have know what was "good practice" on jobs that were done years ago, but personnel continually change in that industry and the people he has today may not know what was done before and will bring to the project their own experience which may not be good.

When someone says they have 25 years of experience in concrete construction, they never qualify it by saying this was good experience or bad experience. As an example, the author recently reviewed a proposal from a very reputable constructor to accept a construction practice which is universally rejected by most building codes and standard practices. In the proposal, documentation was provided which indicated that many years ago, the contractor had convinced an owner to accept the proposed practice based on limited testing by the contractor. This testing had mixed results and was subject to different interpretations. Its implementation was only to make the construction scenario a little simpler for the constructor. This deviation from accepted practice was also for construction in a certain environment. The change in practice impacted the potential for corrosion of prestressing strands. Based on this owners agreement to the change, the next owner was told that the first owner agreed to this so he should also. And he did. The third owner was presented with two prior acceptances and he also agreed. And, it continued with this proposal despite changing structures, materials, and construction environments. When challenged on this, the constructor says "This is the way we've always done it." and that the accepted current practices are "too restrictive".

Whose at fault? Probably all parties, including the owners for not considering the ramifications (risk) of accepting such changes. If the owner had considered the durability aspects of his structure through the use of competent concrete technologists, he probably would never have accepted the change in practice. The need for good construction practices is essential in getting good durability performance. "This is the way we've always done it." is not always the best answer if that way is part of the problem. How can we ensure the constructor will use good durability related practices. It goes back again to education. A good start is to chose a reputable constructor and a proper specification that describes specifically what durability practices are to be implemented. That, by itself, may not be sufficient because there are no guarantees that the specification will be followed. As noted before, a proper QA/QC program is also required.

11 Concluding remarks

The approach to integrating durability design into the structural design process will meet the least resistance if the requirements for durable concrete can be initially included in our national codes and standards. These requirements must be realistic and obtainable. The production of more durable concrete will lead to higher strength concretes. The codes and standards can then undergo future revisions to effectively use the additional strength.

Strong guidance on defining the environment in which the concrete will be used is necessary. This will require more work by the preparers of the codes and standards because current definitions are too general and only consider fairly simple scenarios. To this end, the concrete research community must take a very active role.

A service life requirement should be specified for every structure. Although there can be considerable uncertainty about what a proper service life should, a conservative estimate should always be selected. The incremental costs for improving the life of concrete from 50 to 100 years are infinitesimal compared to the costs of repairing or replacing a structure in the future. The service life selected should, however, be fairly realistic consistent with the planned obsolescence of the structure.

The performance criteria to be included in the codes and standards should focus on the transport properties of the concrete as it is the ingress of the environment into the concrete that is the principal cause of concrete degradation. The necessary criteria generally exists within concrete research community, but must be synthesized and formalized. Prediction of long-term concrete behavior involves the extrapolation of laboratory data, based on the assumption that long term processes, which typically aren't known, will not invalidate the extrapolation. Additional work is still necessary. Because of this uncertainty, the performance criteria should be complimented by some prescriptive criteria which requires the designer to use minimum cement contents, maximum water/binder ratios, supplementary cementing materials, and chemical admixtures. The designer and constructor can not be expected to know what these should be. A strong position must be taken by the preparers of the codes and standards on this point. If we are serious about providing durable concrete, binder contents of less than 400 kg/m^3, water/ binder ratios more than 0.38, and prohibitions on the use of silica fume, fly ash or blast furnace slag are not acceptable for any concrete to be used in an outside environment. Some specifications for specific environmental conditions or classes of concrete have already begun incorporating minimum cement contents or maximum water/binder ratios requirements [1,2,3]. While their requirements are only approaching the suggested limits above, it indicates that this type of solution is recognized by the code writing people. It just needs to be expanded.

The evaluation of performance of the concrete should use national standard tests wherever possible. If there are no such tests, the concrete research community needs to establish and standardize such tests. This may require concessions by many researchers who believe that the method they developed is the only method of merit. Resolution of similar matters like this in other areas of concrete testing has been successfully done and can certainly be achieved for methods of evaluating transport phenomena.

Having done all of the above, there will still be resistance to the concept of providing high performance concrete on a routine basis. The cost of improved durability, from research through demonstration, will be compared with the costs of earlier, less ambitious efforts with the conclusion that the extra costs are not warranted. There is a great need in our industry for life-cycle cost analysis to highlight the need for more durable concrete. And finally, we must realize that durability is a complex topic and compliance with durability requirements may, in some instances, not be sufficient to ensure a durable structure. However, these requirements greatly increase the chances of success. We must also realize that the improvement of durability may not improve the performance of the concrete in other areas. For example, the use of fly ash may reduce early concrete strength development which may be needed for timely construction. There will always be exceptions, but these can be handled on an individual basis.

The above comments, suggestions and recommendations are the views of the author based on his extensive experience in dealing with concrete durability problems in many different regions of the world. Whether you agree with them or not, the need to integrate our knowledge on concrete durability into the design process is real and urgent and steps must be taken as soon as possible to accomplish this. Hopefully, the suggestions offered here are a beginning. The Shilstone [5] have often noted that the many building codes which include guidelines for producing durable concrete based on 28-day strength for selected water-cementitious ratios are the same codes that have guided the industry during the period in history where the durability of concrete has regressed. Either these requirements are not adequate, are not being followed, or are being negated by the use of improper construction practices. A closer look at all these aspects of the durability problem is needed. What we do know is that no concrete <u>yet</u> manufactured will continue to fulfill its role indefinitely. Hopefully, the convening of durability experts at workshops of this type will take another step forward in producing the "eternal concrete".

12 References

1. Canadian Standard S474 (1994), Concrete Structures, Part IV of the Code for the Design, Construction, and Installation of Fixed Offshore Structures, Canadian Standards Association, Rexdale, Ontario, Canada, June 1994.

2. ACI 357R-84 (1984), Guide for the Design and Construction of Fixed Offshore Concrete Structures, Part 4, ACI Manual of Concrete Practice, 1994, American Concrete Institute, Detroit, Michigan.

3. ACI 302.1R-89 (1994), Guide for Concrete Floor and Slab Construction, Part 2, ACI Manual of Concrete Practice, 1994, American Concrete Institute, Detroit, Michigan.

4. Neville, A., "Concrete in the Year 2000," Advances in Concrete Technology, 2nd Edition (Editor: V.M. Malhotra), Canada Centre for Mineral and Energy Technology, Ottawa, Ontario, Canada, 1994, pp 1-17.

5. Shilstone Sr., J.M. and Shilstone Jr., J.M., "Advances in Production Control of Mixtures for High-Performance Concrete," Supplementary Papers, 2nd CANMET/ACI International Symposium on Advances in Concrete Technology, Las Vegas, Nevada, June 11-14, 1995, pp 1-26. (Available from Canada Centre for Mineral and Energy Technology, Ottawa, Ontario, Canada)

3 INTEGRATION OF STRUCTURAL AND DURABILITY DESIGN

T. UEDA and Y. KAKUTA
Hokkaido University, Sapporo, Japan

Abstract
This paper presents a way to integrate structural design and durability design. In the form of the limit state design the causes for deterioration of materials, both concrete and reinforcement, are included in "actions" together with loadings. The deterioration mechanisms are described as material models together with mechanical models. "Analysis" includes analyses for both loading effects and deterioration. "Ultimate and serviceability limit states" are considered as in conventional structural design, but ways to examine the limit states are presented for both loading effects and deterioration. A new concept for serviceability limit states is proposed to consider interaction between limit states caused by different actions. Although examination of limit states caused by the material deterioration may not be practically possible at present, introduction of the integrated design concept will accelerate the necessary study to clarify the deterioration mechanisms.
Keyword: Durability design, limit state design, structural design

1 Introduction

As stock of infrastructures is becoming huge, durability of concrete structures which affects the cost of infrastructures is getting more significant. Studies on durability of concrete structures are currently conducted in various aspects. Some studies emphasize on how to design concrete structures with consideration of durability. Because of lack of information on deterioration mechanisms of materials, both concrete and steel, it is difficult to quantify the deterioration process, such as reduction of cross–sectional area of steel reinforcement due to corrosion. Therefore the way to prescribe provisions for durability design is quite different from that for conventional structural design, and

Integrated Design and Environmental Issues in Concrete Technology. Edited by K. Sakai. Published in 1996 by E & FN Spon, 2–6 Boundary Row, London, SE1 8HN, UK. ISBN 0 419 22180 8.

durability design is usually dealt with in a separate chapter of a design code or in a complete separate code.

Admitting the above mentioned fact, it is tried in this paper to present a possible way to integrate durability design into structural design which is formatted with limit state design concept.

2 Conventional design code

In structural design codes formatted with limit state design concept the following items are seen as main components:

- · Materials
- · Loads
- · Structural analysis
- · Limit states

In a chapter for materials material properties necessary for structural design, such as strengths, creep, shrinkage, relaxation and thermal expansion coefficient, are described. A chapter for loads provides a variety of loads to be considered, such as variable loads, permanent loads and accidental loads as well as how to determine magnitudes of the effects of each load on concrete structures.   Structural analyses, such as linear and nonlinear analyses, static and dynamic analyses, and one–, two– and three–dimensional analyses, are listed as appropriate structural analyses and specified according to limit states.

The main feature of limit state design concept is that clear description of limit states. The definition of limit states is that structural performance beyond limit states becomes worse distinctly.   For example, aesthetics of a structure will be affected beyond a certain crack width, or a structure will collapse beyond a certain amount of deflection.   The former is one of serviceability limit states and the latter is one of ultimate limit states.   Usually limit states are classified into those two limit states. How to examine each limit state by showing an index for a limit state as well as computational process is provided in chapters for limit states.

It cannot be said that durability design has been generally adopted as durability design. Allowable stresses for steel reinforcement in working stress design or a limit state for crack width in limit state design indicate indirectly consideration of effects of steel corrosion on structures. In this sense durability of structures has been considered already. However, there have been some new developments recently.

The latest CEB/FIP Model Code (MC90)[1] includes a chapter for durability which prescribes certain necessary items in design for durable structures. Material models for deterioration mechanism are provided, however, no provision is given on how to use them in design.   Design criteria such as structural form, concrete cover and reinforcement detailing are presented for durability design.

Design codes for durability[2][3] have been proposed with design equations. In Proposed Recommendation on Durability Design for Concrete Structures which was presented by JSCE[2] the following equation is provided for examination of durability:

$$T_p \geq S_p \tag{1}$$

where T_p :   durability index

$T_p = 50 + \Sigma T_p(i,j)$

$T_p(i,j)$:   durability points

S_p :   environmental index

$S_p = S_o + \Sigma \Delta S_p$

S_o :   environmental index for normal environmental conditions

ΔS_p :   increment of environmental index for severe environmental conditions

The durability index indicates how durable a structure is. The greater the durability index is, the more durable is the structure. The durability points represent quantitatively influences of materials, design and construction on durability. For example a higher water–cement ratio or a thinner concrete cover gives a smaller point. No compaction with internal vibrators gives a negative point, but an appropriate protection measure such as coating of concrete gives a positive point. On the other hand the environmental index indicates how severe the environment is to the structure. The greater the environmental index is, the more severe is the environment. The environmental index for normal environmental conditions, S_o is 100 for the case of a maintenance–free period of 50 years and smaller for the cases of shorter maintenance–free period. The increments of the environmental index, ΔS_p are provided for severe salt attack and severe freezing and thawing. Equation (1) can assure the considered maintenance–free period which means a period during which no rehabilitation nor reinforcement works is necessary.

Concept on Durability Design Method of Concrete Structures was proposed by JCI[3]. In the Concept the equation for examination of durability is as follows:

$$C_{de} \leq C \tag{2}$$

where C_{de} :   deterioration depth ($C_{de,n}$ for cases of carbonation, $C_{de,cl}$ for salt attack)

$C_{de,n} = A_n \dfrac{E_n}{Q_n Q_{DC}} \sqrt{t}$

$C_{de,cl} = A_{cl} \dfrac{E_{cl}}{Q_{cl} Q_{DC}} \sqrt{t}$

A_n, A_{cl} :   coefficients

E_n, E_{cl} :   environmental indexes (intensity of action causing deterioration)

Q_n, Q_{cl} :   quality indexes for carbonation and salt attack respectively

Q_{DC} :   quality index for design and construction which is common for carbonation and salt attack

t :   design service life

C :   equivalent concrete cover

$$C = C_0 - C_e + C_a$$

C_0 :   design concrete cover

C_e :   adjustment for construction error

C_a :   increment due to surface finishing material

The limit state corresponding to Eq.(2) is occurrence of corrosion of steel reinforcement. It is mentioned[3] that this limit state may be too conservative, but that there is not sufficient data to predict concrete cover cracking due to corrosion which is more appropriate for a limit state. If Eq.(2) is satisfied, it means that deterioration process due to carbonation or salt attack has not reached the steel reinforcement surface. The deterioration depth depends on cover concrete quality, environmental condition and time considered (the design service life). The equivalent concrete cover is decreased by construction error in the cover and increased by surface finishing.

It should be pointed out that all the existing durability design is mainly for deterioration due to corrosion of steel reinforcement.

3 Outline of integrated design

Current situation of durability design is that it is regarded as a complete separate set from conventional structural design as explained in Chap.2 above. In an integrated design code, however, contents of the current durability design codes will be distributed in corresponding chapters[4] and durability aspects will be considered as limit states as in conventional structural design[4]. The main chapters of the integrated design code are as follows:

- General
- Basis of design
- Materials
- Actions
- Analysis
- Limit states
- Detailing

Material constants and models relating to deterioration which causes durability problems are provided in a chapter on materials. Causes of deterioration are described in a chapter on actions. A chapter on analysis explain how to analyze deterioration process. A chapter on limit states shows how to examine limit states caused by deterioration.

3.1 General and basis of design

Service life of structure is defined in these chapters. Plan for measures, such as rehabilitation and restrengthening, to structural degradation due to actions, including material deterioration is explained. This plan affects durability design greatly.

3.2 Materials
In a chapter on materials, material constants relating to deterioration, such as permeability of concrete, are provided together with mechanical material constants necessary for structural design. Besides material constants, material models necessary for both structural and durability design are prescribed here. Material models for structural design are stress–strain relationships of concrete and reinforcement and mechanical interactions between concrete and reinforcement, while those for durability design are models for transport mechanisms and models for material deterioration such as corrosion of steel and alkali aggregate reaction. For durability design mechanical properties such as stress–strain relationships are also needed, particularly time-dependency of mechanical properties and mechanical properties of deteriorated material. Material properties of surface finishing or coating materials are prescribed in the chapter on materials.

3.3 Actions
In a chapter on actions, all the external actions which affect structural performances should be described. In structural design only physical actions i.e. loadings are considered. However, the chapter on actions includes chemical and biological actions which deteriorate materials as well. In fact chemical actions have been considered in conventional structural design. Requirement for concrete cover and crack width (or reinforcement stress) has been restricted more severely for the cases of structures subject to chemical actions such as sea water and acid gas from factory. In the integrated design, the chemical and biological actions are considered as rationally as physical actions in structural design. For example wave forces considered for off-shore structures are described by equations to indicate their magnitude and frequency for structural design. Chloride contents in sea water and wet–dry cycle due to tide should be described for durability design.

3.4 Analysis
For structural design structural analysis under effects of loadings, using material constants and models are conducted. Methods for structural analysis, such as frame analysis and finite element analysis are specified in this chapter. Durability design also requires structural analysis, but at the same time deterioration of material characteristics should be considered. Deterioration of material characteristics, such as reduction in effective area of steel due to corrosion and in strength of concrete due to alkali aggregate reaction, which is generally time dependent, is predicted by appropriate analysis. The analysis may be linear or nonlinear, one-, two- or three–dimensional. Finite element method can be applied for the analysis.

3.5 Limit states
Generally ultimate limit states and serviceability limit states are considered in limit state design. Ultimate limit states are related to safety, while serviceability limit states are related to serviceability. The limit states, usually ultimate limit states, caused by

fatigue and accidental loadings (usually earthquake) are often treated separately because of difference in magnitude of loadings. The magnitude of loadings for fatigue is significantly less than that of ordinary loadings which cause ultimate limit states. That for earthquake is, however, greater. Considering the difference in the nature of actions, the limit states caused by deterioration of materials can be treated separately like those due to fatigue and earthquake. Limit states under effects of material deterioration possibly occur under less magnitude of ordinary loadings, even under no effect of ordinary loadings.

Limit states themselves may not necessarily be different according to actions: ordinary loadings, fatigue, earthquake and durability related actions. Failure of members is a typical ultimate limit state and a certain concrete crack width is a typical serviceability limit state under any effects of actions. Examples of limit states caused by material deterioration are member failure in tension due to break of reinforcement whose cross-sectional area is reduced due to corrosion and concrete crack opening due to swelling of reinforcement caused by corrosion. Suggested contents for a chapter on limit states are as follows:

· Serviceability limit states
> Classification according to indexes:
>> Concrete crack width (for comfort of users, aesthetics and water/air tightness)
>> Deflection (for comfort of users)
>> Vibration (for comfort of users)
>> Concrete surface appearance (for aesthetics)
> Classification according to actions:
>> Serviceability limit states for ordinary loadings
>> Serviceability limit states for earthquake (or accidental loadings)
>> Serviceability limit states for fatigue
>> Serviceability limit states for durability

· Ultimate limit states
> Classification according to indexes:
>> Member load carrying capacity (in axial force, flexure, shear and torsion)
>> Deflection (in axial force, flexure, shear and torsion)
>> Collapse or instability of structure (buckling, falling, sliding, etc.)
> Classification according to actions:
>> Ultimate limit states for ordinary loadings
>> Ultimate limit states for earthquake (or accidental loadings)
>> Ultimate limit states for fatigue
>> Ultimate limit states for durability

The serviceability limit state of concrete crack width is examined as in structural design by the following equation:

$$\frac{w_{cr,d}}{w_{cr,a}} \leq 1.0 \tag{3}$$

where $w_{cr,d}$: factored crack width caused by actions

  $w_{cr,a}$: factored limit for crack width

The factored crack width is a calculated crack width caused by actions, that is ordinary loadings, earthquake, fatigue or durability related actions. The calculation is based on the material properties provided in Sec.3.2 and the analytical method presented in Sec.3.4. How to calculate the factored crack width is different for each action, as a result the calculated values are different. Values for the factored limit for crack width are chosen according to the nature of limit states in terms of comfort of users, aesthetics or water/air tightness.

Similarly the ultimate limit state of member load carrying capacity is examined by Eq.(4).

$$\frac{S_d}{R_d} \le 1.0 \tag{4}$$

where S_d :   factored force caused by actions

  R_d :   factored member load carrying capacity

The factored forces are calculated with the analytical methods in Sec.3.4 using the material properties in Sec.3.2. The calculated results are different among the different actions. The factored member load carrying capacities are also calculated with the material properties in Sec.3.2. The material properties may be different for different actions, so that the calculated member load carrying capacities may be different. For example the flexural capacity for ordinary loadings is likely to be greater than that for durability related actions because decrease in material strength is caused by material deterioration.

3.6 Detailing

If all the necessary limit states described in Sec.3.5 can be examined appropriately, provisions for detailing are not required. However, some limit states may not be examined because of lack of information. In such cases proper detailing may be required to assure minimum structural performance. Some detailing to prevent unexpected or unacceptable material deterioration may be necessary.

Proper detailing can simplify the analysis. For example minimum reinforcement can maintain crack width within a certain limit, so that even a simple design calculation will be able to predict the extent of reinforcement corrosion accurately.

4 Interaction between limit states caused by different actions

Among actions of ordinary and accidental loadings, a limit state caused by one action is generally not affected by the others. Ultimate limit states caused by an ordinary loading with extremely large magnitude, such as member failure in flexure, are not affected practically by other ordinary loadings having already acted to the structure.

On the other hand, effects due to all ordinary loadings having already acted are considered for fatigue limit states.

Limit states due to material deterioration are very sensitive to effects of not only actions causing material deterioration directly, e.g. chemical and biological actions, but also other actions. If an ordinary loading causes cracking in concrete, corrosion of steel reinforcement or alkali aggregate reaction of concrete may be affected significantly. Easy absorption of oxygen and water into reinforcement surface through cracks may accelerate corrosion. In this sense crack width or reinforcement stress is limited in conventional structural design. The limit states such as cracking (or crack width) due to corrosion and reinforcement break due to corrosion, therefore, should be examined with consideration of effects of ordinary loadings.

Instead of consideration of interaction between effects of different actions, another limit state for each action, in which limit states caused by other actions will not be affected practically, may be introduced. For example, a serviceability limit state in which no concrete cracking takes place under effects of ordinary loadings can be chosen like in prestressed concrete structures. Or during moderate earthquake, concrete crack width is limited, so that significant effect of the cracks on material deterioration can be avoided.

5 Example of design process

An example of design process is illustrated in this chapter. The considered limit state is a serviceability limit state of crack width for aesthetics. The actions considered are durability related actions which cause corrosion of steel reinforcement.

The considered environmental conditions are described by actions in Sec.3.3. Transport of corrosive factors, such as chloride and oxygen, and carbonation in cover concrete after the considered service life of the structure is estimated, using the material properties in Sec.3.2 and the analytical methods in Sec.3.4. The amount of corrosion is then calculated under the estimated transport and carbonation using the material properties in Sec.3.2. If cracking due to other actions takes place in the cover concrete, the effect of cracking on the amount of corrosion is considered. Crack width under the calculated amount of corrosion is calculated based on the material properties in Sec.3.2. The calculated crack width is compared with the limit of crack width for aesthetics. If the calculated crack width is greater than the limit, the structure should be redesigned.

6 Future works towards implementation of integrated design

Introduction of limit state design concept to structural design has been promoting studies to clarify mechanical behavior of concrete structures in order to examine limit states more reasonably. At present deterioration mechanisms are not sufficiently clarified. Introduction of the integrated design concept would be expected to help identify lacking information and accelerate studies to clarify the deterioration mechanisms.

The integrated design can indicate clearly when and which maintenance works will be needed and allow engineers to calculate rather easily the life time cost of structures. Appropriate concepts on the life time cost in stead of the initial cost should be developed and implemented in design codes. The life time cost, however, depends on non–engineering factors, such as economical situation and society demands upon structures. Future cost, such as maintenance cost, significance of structures to society in the future, and life span of the structures according to society demands may not or should not be predicted by engineers. Role of engineers, therefore, is to provide society with necessary information on engineering matters, so that more precise life time cost can be predicted.

7 Concluding remarks

A way to integrate structural design and durability design which are currently provided separately is presented. Material characteristics relating to its deterioration are included together with material mechanical characteristics. Causes of material deterioration are prescribed as actions together with loadings. Serviceability and ultimate limit states caused by material deterioration are prescribed. A new concept for serviceability limit states is proposed to consider interaction between limit states caused by different actions. It is expected that more rational design can be achieved through the integrated design. Although examination of limit states caused by material deterioration may not be practically possible at present, introduction of the integrated design concept will accelerate the necessary study to clarify the deterioration mechanisms.

8   References

1.   Comite Euro–International Du Beton. (1993) *CEB–FIP Model Code 1990*, Thomas Telford, London.
2.   JSCE Subcommittee on Durability Design for Concrete Structures. (1989) *Proposed Recommendation on Durability Design for Concrete Structures*, Japan Society of Civil Engineers, Tokyo.
3.   JCI Committee for the Research of Durability Design Method of Concrete Structures. (1991) *Concept on Durability Design Method of Concrete Structures*, Japan Concrete Institute, Tokyo (in Japanese).
4.   Villas, J.M. (1990) *A Study on the Durability Design of Concrete Structures*, Master's thesis submitted to the Asian Institute of Technology, No. ST–90–24, Asian Institute of Technology, Bangkok.

4 GENERAL CONCEPT BRIDGING BETWEEN STRUCTURAL DESIGN AND DURABILITY DESIGN

K. MARUYAMA
Department of Civil Engineering, Nagaoka University of Technology, Nagaoka, Japan

Abstract
Service life is an essential concept for the design of structures. Most design codes involve the concept implicitly, but many of them does not tell how the   life span is assured. Durability design treats how materials are getting deteriorated and gives some instructions on how to keep the materials in sound condition within the service period. Combining the durability design with the structural design, we can make the design of structure effective within a given service life.
Keywords: Deterioration, durability design, general concept, service life, structural design

1   Introduction

The history of making structures is, no doubt, very long and the structural design has also a long history with or without written documents. In this century the structural design methods have been remarkably progressing with a lot of findings in structural behaviors. Recent development in electronics, in particular in the field of computer science, makes it quite easy to analyze the structural behaviors in detail by numerical computation.

On the other hand, the concept of durability is relatively very new. It has been only twenty or thirty years since structural engineers recognized that concrete structures could not survive forever. Some causes for rapid deterioration of concrete structures may attribute to poor construction practice, but in other aspects inevitable reasons exist causing reduction of the service life due to expansion of our activities. We are able to and have to construct many structures under severe conditions, such as marine

Integrated Design and Environmental Issues in Concrete Technology. Edited by K. Sakai. Published in 1996 by E & FN Spon, 2–6 Boundary Row, London, SE1 8HN, UK. ISBN 0 419 22180 8.

environment. This forces us to face the durability of concrete structures.

In these twenty years, we have seen many concrete structures in the coastal region of Japan See suffering from salt attack, with their service lives being considerably reduced. Similar problems have been reported in the Okinawa Island in Japan. A new design concept for structures is urgently requested.

2   JSCE code on structural design [1]

Japan Society of Civil Engineers (JSCE) has adopted the limit states design method for the structural design since 1986. In the JSCE code three limit states are specified as the serviceability limit state, the ultimate limit state and the fatigue limit state. As for examination of each limit state, five safety factors are introduced, such as a material factor, a member factor, a load factor, an analysis factor and an important factor. Any safety factor, however, does not count in deterioration.

The code describes that a designer, first of all, should specify the service life of structure. But it says nothing on how to assure the life span. The fatigue limit state may correspond directly to the service life only when repeated loadings are influential on structures. The strength deterioration of materials to be taken, however, is only due to the repetition of loading, not due to any chemical reaction during the service life.

The serviceability limit state is partly concerned with durability. The code describes how to examine the flexural crack widths of reinforced and prestressed concrete members. The difference of environmental condition is taken into account for the specification of the allowable crack widths. The environment is categorized into three, such as ordinary, severe and extra severe conditions. It should be noticed that the allowable crack widths are not given as any definite values, but a function of cover thickness. A larger cover thickness must be better to protect steel reinforcement from corrosion although it certainly results in a larger crack width at the surface of concrete. The material deterioration, again, is not considered in this section.

3   JSCE recommendation on durability design [2]

The JSCE recommendation on durability design must be very unique in the style as well as the content. It adopts a point system to evaluate how materials, design details and construction practice are appropriate for durability. The severity of environment is also evaluated as points. The uniqueness of this method lies in the way to express the durability by summarizing all the points of materials, design details and construction practice on the equal base. The evaluation is done by examining the total point of durability being in excess of the environmental point.

Materials are examined in many aspects, such as types of cement, soundness of aggregates, mix proportion (unit contents of cement and water, W/C, slump, admixture, etc.). Design details on concrete cover thickness and spacing of reinforcing bars are most important for durability. Construction practice is categorized in many items including the effect of workmanship. Each item is weighed according to the relative intensity for durability.

This method, however, does not count in the deterioration process of structures. The recommendation says that the method may only assure concrete structures free from maintenance for 50 to 100 years under a given environmental condition.

4 Assurance of service life

It must be the final goal in the durability design to predict and assure how long concrete structures survive under given conditions. As for concrete, many studies have been carried out for carbonation, neutralization, penetration and accumulation of chloride [3]. We have some knowledge on how fast such chemical processes go. We also have some idea on how steel reinforcement is corroded. This knowledge, however, is not enough to predict when concrete structures come to the end of their service life, or how long the structures can survive. Much more studies are required on the deterioration of stress transfer mechanism, such as bond deterioration.

Studies on the deterioration process are very time-consuming as no effective acceleration test method exists. To obtain reliable results, we have to use an actual time lapse. However, without study of the deterioration process, we can not assure the service life of structures.

5 Behavior beyond service life

Deterioration process, in general, is slow at the first stage, and is accelerated beyond a certain point. This may be called the second stage. At the final stage structures lose their capacity considerably. From this point of view, the service life should be set within the first stage of deterioration. However, we may have to use structures continuously beyond the service life for some reason in the future. Then, the structural behaviors should be studied all the way through the deterioration process. Proper maintenance and repair could reduce the deterioration speed. Some strengthening methods may retrieve the functions of structure up to a satisfactory level.

For example, the deterioration due to salt attack appeares as described in the following;

(1) Penetration and accumulation of chloride ion in concrete: no visible deterioration.
(2) Recognition of rust colored spot on the surface of concrete: start of rusting of steel bars.
(3) Appearance of narrow longitudinal crack: this stage may be defined as being beyond the service life. The mechanical properties of structure, however, are still in acceptable conditions.
(4) Spalling off of concrete cover: bond is almost lost whereas reduction of cross sectional area of steel bar may not be large. Retrieval of capacity and function may be possible by repair and strengthening.

In the case of salt attack, the key lies in how much the cross sectional area of steel bar is reduced. In general, it is very difficult to prevent the invasion of chloride ion, but from the surface observation it is relatively easy to detect the rusting of steel bar in concrete at the early stage. Then, repair and strengthening may be conducted in time before collapse of the structures.

6   Bridging between structural design and durability design

In the current JSCE design methods, the structural design is rather independent to durability. The durability design is still in the qualitative stage although every item in the design is evaluated quantitatively in points. In order to bridge between the two design methods, the the deterioration process of structures must be made clear in terms of stress transfer mechanism as well as materials.

Once the deterioration process is made clear, then the durability considerations are explicitly introduced into the structural design in the following manners;

(1) Service life: we may determine the service period without maintenance and the extra life with proper maintenance and repair.

(2) Material factor: we may choose proper values with consideration of strength deterioration.

(3) Member factor: this should reflect the bond deterioration.

(4) Crack control: in combination eith the concrete cover thickness, the allowable crack widths should be related to the service life.

(5) Design details: concrete cover thickness is determined with consideration of service life. Surface coating may be taken into account. The minimum reinforcement ratio and the minimum diameter of reinforcement should be influenced by the service life.

It should be noted that important infrastructures will often continue to be used even beyond the original service life. For such a case the way of maintenance and repair must be considered in the design process.

7   References

1.   Japan Society of Civil Engineers. (1988) *Standard Specification for Design and Construction of Concrete Structures - 1986. Part 1:Design*. JSCE, Tokyo. Concrete Library International, No.12.

2.   Japan Society of Civil Engineers. (1989) *Proposed Recommendation on Durability Design for Concrete Structures*. JSCE, Tokyo. Concrete Library International, No.14.

3.   Japan Concrete Institute. (1991) *Consideration on Durability Design for Concrete Structures*. JCI, Tokyo. JCI-C24. (in Japanese)

5 PREDICTION OF DURABILITY OF CONCRETE: RESEARCH NEEDS

S.P. SHAH
NSF Center for Advanced Cement-Based Materials,
Northwestern University, Evanston, Illinois, USA

Abstract
Concrete for the twenty first century can be much stronger, more durable and at the same time cost and energy efficient. However, this will not be possible unless we understand this material better. Many of the problems that exist in concrete structures have resulted from lack of knowledge about factors affecting durability of concrete as well as failure to apply existing knowledge. The durability of concrete is defined as the ability of concrete to withstand destructive action of weathering or service environment. One of the most critical factors influencing durability of concrete is its permeability or ability to resist penetration of such substances as liquids, chloride ions, sulfate ions, or other aggressive chemicals. In order to predict concrete durability one has to understand concrete microstructure, transport properties, microcracking and their interaction. Some aspects of concrete durability and some of the future research needs are presented.
Keywords: Concrete, cracking, durability, microstructure, permeability, porosity, shrinkage.

1 Relating transport properties with microstructure

Concrete durability is closely linked to its microstructure, more specifically to its permeability. Permeability of concrete is related to its pore structure, porosity, pore size, and tortuosity. To improve durability we must understand the characteristics of pore structure, diffusion and transport properties of concrete.

1.1 Chloride permeability
The water permeability tests which are based directly on Darcy's Law are difficult to perform on concrete and do not yield a reliable results. This is mostly because a great

Integrated Design and Environmental Issues in Concrete Technology. Edited by K. Sakai. Published in 1996 by E & FN Spon, 2–6 Boundary Row, London, SE1 8HN, UK. ISBN 0 419 22180 8.

amount of time is required to achieve the equilibrium conditions, and extremely low flow rates make an accurate measurements of outflow very difficult. Because of these difficulties to this day no standard ASTM permeability test procedure based directly on Darcy's Law has been developed. In order to avoid some of the difficulties associated with complex water flow tests, rapid chloride permeability test (ASTM C-1202) of concrete has been developed. The principle of this rapid chloride permeability test procedure is to determine the amount of electrical current passed through the concrete sample exposed to the potential of an electric field over a fixed period of time. Since the rapid chloride permeability test has its basis in electro-osmosis and measures chloride ions diffusion under the action of an electric field, the test requires an electrolytic "path" through the sample. As the electrical resistance of this path will be influenced by the moisture content of the sample it is important to provide a full saturation of the sample prior to testing.

There are several reasons concrete could exhibit artificially high chloride permeability. Many additional ions are introduced into the concrete in the various admixtures. This may cause the concrete to be more electrically conductive and therefore make it appear to be more permeable than it really is. This suggests that the rapid chloride permeability test should be used with caution for any concrete which has excess ions. Further research is needed to relate chloride permeability to microstructure and permeability.

1.2   Impedance spectroscopy

A unique feature of the microstructure of a cementitious matrix is the presence of a well connected and finely divided pore system. The continuously evolving network of pores governs the physical properties of cement-based materials during hydration, as well as the ultimate physical properties, including strength, permeability, shrinkage, and creep. Cement paste is electrically conductive by virtue of its interconnected pore network filled with water containing mobile ions. Therefore, electrical properties, most notably conductance, have been long employed to study the evolving pore structure. Electrical property measurements have distinct advantages over other microstructure-sensitive techniques such as mercury intrusion porosimetry, and electron microscopy. Electrical property measurements are noninvasive and nondestructive and can sample the microstructure in-situ without recourse to water removal, usually by heating, which is known to permanently alter paste microstructure. The two major disadvantages of electrical property measurements have been the inability to differentiate between the underlying factors responsible for conduction (e.g., ion concentration, water content, pore structure) and the lack of adequate quantitative models to relate the electrical properties to the microstructure. Recently, impedance spectroscopy has been applied to the study of cement-based materials. Impedance spectroscopy involves measuring complex electrical properties of cement when current with frequency varying from DC to Gigahertz is applied to a concrete sample. It has found that with this method we can measure how microstructure changes as cement hydrates. Impedance Spectroscopy also allows for prediction of important engineering properties such as permeability and diffusivity. Right now impedance spectroscopy is only used for cement paste, further research is needed to introduce the technique to predict concrete permeability.

1.3   Digital based computer model

Cement-based materials are random composite materials over many length scales. In common with any random material, the two main problems that must be solved to obtain quantitative theoretical understanding of microstructure-property relationships are: (1) what actually is the microstructure and how it can be accurately modelled, and (2) given enough knowledge of a microstructure, how can properties be calculated?

Portland cement is a complex particulate system consisting of multisize, multiphase particles. With the large memories and fast computational speeds of late 20th -century computers, it is now possible to attack problems, namely random grain problems like those of cement based materials. For these kind of materials, the important randomness for most problems is at the grain or particle level, not the atomic level, and is complicated by the fact that their microstructure changes markedly on time scales of hours and days due to hydration reactions. Computational materials science techniques using digital-image-based microstructure development models can adequately simulate many features of the microstructure of cement paste, mortar and concrete. This is evidenced by the agreement between computed properties for simulated microstructures and available experimental measurements, specifically conductivity and permeability. These kinds of simulations are providing for the first time, quantitative understanding of microstructure-property relationships for cement-based materials.

1.4 Statistical model

Statistical modeling can also be developed to relate transport properties of concrete, namely chloride permeability to structural characteristics. This method involves reconstructing the three-dimensional structure of concrete. A well defined aggregate dimension (e.g. spherical aggregates) is helpful because the reconstruction can be easily achieved. The full reconstruction may be an intermediate step, if we can develop accurate stereological procedures applied to the two-dimensional slice faces. The arrangement of aggregate particles (connectedness and clustering) should affect chloride permeability. Since chloride ions are believed to flow most readily through the interfacial zone between the aggregate particles and cement paste, it may be possible to characterize this path of preferential flow.

2 Probabilistic research

One of the major problem associated with the durability of concrete is the development of proper test methods. Most of the tests performed in the laboratory are accelerated tests, since it is difficult to conduct tests over an extended period of time. It is thus essential to properly interpret these results to assess the long-term durability of real life structures. Concrete is a non-homogenous material, which makes it more difficult to predict its durability. Even for same concrete in similar environmental conditions the performance is sometimes different. As a result, probabilistic methods are becoming increasingly popular as well as necessary for this purpose.

3 Thermal and shrinkage cracking

Concrete shrinks when it is exposed to a drying environment. The amount of shrinkage depends on many factors including the properties of the material, temperature, relative humidity of the environment, the age when concrete is exposed to drying environment and the size of the structure. If the concrete is restrained, then tensile stresses may develop due to shrinkage and cracking occurs. This cracking is a major concern for flat structures like highway pavements, industrial floors, parking garages and bridge decks. The development of crack width with time depends on many factors such as free shrinkage, creep, elastic modulus, tensile strength and fracture toughness. With the use of increased cement content, silica fume and water reducers to achieve high strength concrete, thermal as well as shrinkage cracking are getting more attention. Thermal and shrinkage cracking performance of high strength concrete may be substantially poor due to higher chemical shrinkage, higher rate of heat evolution, lower specific creep, higher modulus of elasticity and lower aggregate-interlock along the cracked faces.

To evaluate performance of different material composition, we need to develop appropriate tests to measure cracking due to thermal and restrained shrinkage stresses and develop predictive models. A considerable amount of research is currently underway on these subjects.

3.1   Test to evaluate shrinkage cracking
Free shrinkage tests alone may not offer sufficient information on the behavior of concrete structures since virtually all concretes are restrained in some way, either by reinforcement or by the structure. Thus restrained shrinkage tests have been developed to measure the shrinkage cracking behavior of concrete. Test methods commonly used for measuring shrinkage cracking of concrete are: (a) bar test; (b) plate test; and (c) ring test   The bar test suffers from several disadvantages, in particular the difficulty of providing a constant restraint. The advantages are, however, uniaxial stress and possibility of using large aggregate size. The plate provides a biaxial restraint which depends on geometry and boundary conditions. The primary difficulty with the plate test is to estimate the actual extent of restraint provided by the stirrups. Ring test provides a high and nearly constant restraint, which enables to test cement paste, mortar and concrete. Because of the axisymmetry, the geometry and boundaries do not significantly influence the result. Since the free shrinkage of bar and ring specimen has been shown to be equal, especially in early ages, this test method has considerable promise. Research is also underway to develop test methods for measuring thermal cracking.

3.2   Modeling
The physical concept of the formation of cracks under restrained conditions can be described as follows. At a given time, the composite undergoes a certain shrinkage strain $\epsilon_{sh}(t)$. If the movement needed for this shrinkage is prevented by restraint, then a tensile stress $\sigma(t)$ is developed. This tensile stress $\sigma(t)$, which is a function of both strain $\epsilon_{sh}(t)$ and the elastic modulus of the composite $E(t)$ at time t, can be expressed as

$$\sigma(t) = E(t)\, \epsilon_{sh}(t) - \sigma^R(t)$$

where $\sigma^R(t)$ is the stress reduction provided by relaxation effects. Note that stress strain, and elastic modulus are time-dependent variables. When the resulting stress $\sigma(t)$ reaches the tensile strength of the composite at t, then a crack will form. After the crack forms, the uncracked portion will continue to shrink, resulting in a widening of the crack. In order to accurately predict shrinkage cracking one has to have time dependent values of tensile strength (more realistically fracture toughness), elastic modulus, creep, free shrinkage and autogenous shrinkage. It is also necessary to incorporate the effect of stresses built up by thermal variations and temperature gradient over the thickness of the specimen.

A realistic method to incorporate effects of boundary constraint, gradients in humidity and shrinkage, and the concepts of fracture mechanics need to be developed.

4 Interaction between microcracking and transport properties

Much of the literature available on permeability deal with uncracked concrete. The permeability tests are generally carried out on uncracked and unloaded specimens. Not much research has been performed on permeability of cracked concrete. From the durability view point a structural element should also work as a barrier against water pollutants. The concrete's tightness can be guaranteed for example by avoiding crack formation and the development of small and fine cracks or limiting the crack depth. A clear understanding of the interaction between microcracking and transport properties is essential to make concrete structures more durable and economical.

5 Concluding remarks

To improve durability of concrete structures we need to be able to predict durability, that is related material composition with the various aspects of durability such as chloride attack, freeze-thaw resistance and alkali-aggregate reaction. Accurate methods of assessing durability involves atleast following steps: (1) relate microstructure to transport properties, (2) develop appropriate performance tests, (3) interaction between stresses and durability, (4) concepts of probability, and (5) methods of monitoring degradation in the field. A substantial amount of new information is needed before we rationally design structures for a prescribed durability.

6 References

The following references provide a more detailed information about some of the research summarized in this paper.

1. Halperin, W.P., Jehng, J., Song, Y. (1993) Application of Spin-Spin Relaxation to Measurement of Surface Area and Pore Size Distributions in a Hydrating Cement Paste, Magnetic Resonance Imaging, Vol. 12, pp. 169-173.
2. Christensen, B.J., Mason, T.O., Jennings, H.M. (1992) Influence of Silica Fume on the Early Hydration of Portland Cements Using Impedance Spectroscopy, Journal of the American Ceramic Society, Vol. 75, No. 4, pp. 939-945.
3. Christensen, B.J., Coverdale, R.T., Olson, R.A., Ford, J.F., Gorboczi, E.J., Jennings, H.M. and Mason, T.O. (1994) Impedance Spectroscopy of Hydrating Cement-Based Materials: Measurement, Interpretation, and Application, Journal of American

Ceramic Society, Vol. 77, No. 11, pp. 2789-2804.

4. Grzybowski, M. and Shah, S.P. (1990) Shrinkage cracking of fiber reinforced concrete, ACI Materials Journal, Vol. 87, No. 2, pp 138-148.

5. Sarigaphuti, S., Shah, S.P. and Vinson, K.D. (1993) Shrinkage cracking and durability characteristics of cellulose fiber reinforced concrete, ACI Materials Journal, Vol. 90, No. 4, pp 309-318.

6. Shah, S.P. and Ouyang, C. (1991) Mechanical behavior of fiber-reinforced cement-based composites, Journal of American Ceramic Society, Vol. 74, No. 11, pp 2727-2953.

7. Marikunte, S. and Shah, S.P. (1994) Engineering of cement-based composites, Concrete Technology: New Trends, Industrial Applications, Proceedings, The International RILEM Workshop, E & FN Spon, London, pp 83-102.

6 SERVICE LIFE DESIGN FOR THE NEXT CENTURY

S. ROSTAM
COWI, Consulting Engineers and Planners AS, Lyngby, Denmark

Abstract
Classical procedures for design, construction and use of concrete structures have failed to provide reliable long-term performance of structures exposed to aggressive environments. An operational approach to improve the situation will require a multi-disciplinary input from educational, technical, economical and political experts. This paper outlines the interacting technical and non-technical elements of design for durability and performance which are necessary to obtain a satisfactory service life design.
Keywords: Build environment, communication, construction, deterioration mechanisms, education, responsibilities, service life design, uncertainty.

1 Durability: theory versus practice

Much valuable knowledge on factors which influence the durability of concrete structures is available within the research community. Vast experience on how to ensure long term quality through appropriate execution is available in the design and construction industries. However, we still produce bad concrete and execute structures with durability problems. Why?

The answer is complex, of course. If it was easy, we would have solved the problems long ago.

At national and international conferences very valuable research results and observations from practice are presented, but the advice apparently falls on stony ground and the message seems not to reach the design offices and construction sites. We must face up to the fact that the knowledge and experience we have already gained is not adequately reflected in many new concrete structures.

Integrated Design and Environmental Issues in Concrete Technology. Edited by K. Sakai. Published in 1996 by E & FN Spon, 2–6 Boundary Row, London, SE1 8HN, UK. ISBN 0 419 22180 8.

This illustrates the prevailing serious gap between research and practice. We are faced with communication problems which have to be overcome urgently in order to achieve real improvement in the long term performance of concrete structures.

The people involved represent the four functions commented on below. These widely ranging fields of interests must communicate their requirements and calibrate their performance more effectively if improvements are to be expected, [1]. This illustrates the complexity of the building and construction sector.

- Engineering formation
  First we have to look at the engineering formation. At many universities, the multi-disciplinary character of the problems which have to be solved has not been acknowledged to a satisfactory degree. Polytechnical engineering competence has been lost during the past few decades for the benefit of specialisation.

- Construction engineering
  Once out of university, the construction engineer will not usually be present at conferences and seminars, even though first-hand eye-opening information on experience with advanced concrete technology may well be picked up by the observant participant. This could, in many cases, trigger innovative developments for practice, but more often than not this opportunity is lost. The research engineer also often fails to pursue his research achievements all the way to this practical application. All in all, today's traditions within construction engineering make it is difficult to ensure site use of new knowledge becoming available.

- Design engineering
  The design engineer is really the person expected to bridge the gap between research and practice. However, he is usually squeezed between tight budgets and tight time schedules on the one hand and his inherent desire to keep up with rapid developments in technology. This limits his possibilities to meet expectations fully.

- Ownership
  As the construction industry is governed by competition, there is no motivation for providing anything better than the client demands - and what he pays for. Economic requirements - or rather economic restraints - usually dominate the technical requirements. Often the owner has difficulties in overviewing the long term consequences of alternative decisions. Also, the engineer may have difficulties in providing factual documentation of the real costs of future repair and maintenance works associated with alternative solutions. One reason is that important decisions outside the hands of the engineer, and perhaps also outside the hands of the individual owner, may have decisive influence on which alternative may become the optimal one in the future. Such decisions are often of a political nature. Hence, short term optimization dominates. This has lead to a low-bid syndrome prevailing for the large majority of construction works. This does not support initiatives to provide long term durable solutions, as it often - though not always - means slightly increased initial costs.

1.1 Information and education

When a technology is undergoing serious changes and experiences rapid development, the gap between frontline knowledge and everyday practice becomes larger and larger. This is one of the disturbing facts of today's interaction between modern concrete technology and construction practice. Considerable national and regional differences do not conceal the fact that this is a global problem for our profession. The most serious consequence is enormous losses to society in both short and long term.

To bridge the most serious communication gaps, intensified and modified forms of dialogue are needed.

The first step towards improvement is to educate the owners of structures to understand the basic decisions which govern the service life of their concrete structures. To make them see both the short-term and long-term economic consequences of alternative decisions.

Public owners are the most important because they show the way ahead. However, they are to a large extent guided by political strategies. The true target for more detailed information should be the real decision makers, and in many cases they have practically no knowledge of engineering. These decision makers may be the politicians whose horizons are very near to hand, ruled by election dates or the media, often guided only by the situation of today and ignoring the long-term consequences of decisions which do not sell papers nor attract viewers.

We, the engineering community, have failed to communicate our message to the right decision makers, regardless of how obviously beneficial we regard long-term service life strategies to be for society.

The next step is to establish a more penetrating dialogue between the construction industry and the universities. In addition to the much-needed fundamental research, performed under full autonomy by the universities, applied research and development need stronger input and participation from industry in order to push the sector forward. In several countries, this direct link between industry and university is very weak or completely lacking, which in the long run will punish the domestic construction industry on two levels. One is the growing lack of updated and competitive technical input needed by the industry, the other is the production of future engineers who are not being given a fully relevant education. Many universities could be accused of having failed in this respect.

As a third step, the contractors too must increase their awareness. Whenever they are represented at concrete conferences, they must be made aware of their responsibility to pass on the information they have acquired and to supervise on site concrete works.

In contrast to the problems described above, the technical problems in creating durable structures are surprisingly simple. We "just" have to consider a maximum of three basically different types of deterioration mechanisms, all governed by only three transport mechanisms as described in Section 4.

When we can model these mechanisms at the design stage and control them to a reasonable degree during construction, we have reduced the durability problems to a fraction of what they are today. But the verification of qualities achieved must be carried out on the real structures, not in the laboratory alone, as is mostly done today.

2   Durability technology

To ensure long-term durability for concrete structures, the events which threaten their durability must be identified. It must also be understood how the structures react to these events. This means that the aggressivity of the environment and the possible deterioration mechanisms should be identified at the design stage.

The design of durable structures with a correct performance will then have to concentrate on two parallel activities:

- ensuring adequate resistance to the predicted external environmental effects
- providing satisfactory load-carrying capacity and safety under the expected loadings.

To provide adequate resistance against aggressive environmental actions, it is necessary to understand how reinforced concrete structures deteriorate to know how such deterioration is prevented or, at least, to ensure sufficiently slow deterioration.

The key to creating durable structures is the availability of models of the deterioration mechanisms. Thereby the criticality and sensitivity of different parameters can be evaluated prior to choosing solutions which can provide sufficiently durable structures. The corresponding technical discipline is called "durability technology".

A rational engineering approach to ensuring durable structures is to make use of scientifically sound models which describe the materials' reactions to different types of environment. The benefit of such modelling is the theme running through the modern approach to design durable concrete structures, [2,3,4].

3   Environmental aggressivity

To define the aggressivity in which the structure is to be placed becomes an essential part of service life design. Unfortunately, classifying environmental aggressivity is the weakest link in the chain of decisions needed to provide long term durable structures. In particular, the identification of the micro-environment based on macro-environmental observations is lacking. The direct interaction between the environment closest to the surface of the structure is the most important of these factors.

Current European activities concentrate on establishing generally applicable definitions which relate directly to the individual deterioration and transport mechanisms. Previously, very simplified definitions were used, mainly based on the macro-environment and independent of the deterioration mechanisms. The new approach relates more directly to the micro-environment and holds out much more promise.

4   Deterioration of concrete structures

In practice, the number of really significant deterioration mechanisms is limited. There are really only three basic mechanisms to consider:

- physical deterioration, e.g. freeze-thaw action including de-icing salts and the restraining effects of steep thermal gradients, cracking, abrasion and wear, or salt scaling due to re-crystallisation of salt in the pores where the expansive pressure of crystal growth leads to surface scaling of concrete of poor, or mediocre, quality and strength.

- chemical deterioration, i.e. concrete reacting with the surrounding media causing either expansion of the concrete leading to cracking and spalling (alkali-silica reactions (ASR) or sulphate attack), or dissolution and disintegration of the cement paste which binds the fine and coarse aggregates together.
- electro-chemical deterioration, i.e. corrosion of steel reinforcement having been depassivated either by carbonation or chloride contamination of the surrounding concrete, or a combination of both. Rust expansion causes cracking and spalling of the concrete. With restricted access of oxygen, such as under very moist conditions, black rust with very limited expansion may occur, leading to steel section reduction without the valuable warnings of cracks and spalling.

4.1   Governing parameters

Water and salt are among the most aggressive substances threatening the durability of concrete structures. In fact, no serious deterioration takes place without sufficient moisture or water available. Any attempt to reduce the moisture exposure of structures in the atmosphere will have beneficial effects on the service life.

Chloride-based salts incur a serious risk of local pit corrosion of the bars when the chlorides reach the reinforcement in sufficient quantities. If the salt contains alkali-metal ions, they will enter the concrete and increase the total alkali content, with increased risk of alkali-aggregate reactions if the concrete contains reactive aggregates.

As a de-icing agent, salt causes a freezing shock to the surface layers when the ice is forced to thaw. The results may be temperature-induced scaling, delamination and crumbling of the concrete surface.

Salt is hygroscopic. It retains water, which makes it difficult to dry up salt contaminated concrete.

The rate at which deterioration takes place is strongly influenced by the temperature. A simple rule of thumb says that an increase of 10 °C in temperature will double the rate of chemical and electro-chemical reactions. Hence marine structures in hot environments are among the most seriously exposed concrete structures.

4.2   Transport mechanisms

The type and rate of deterioration depend on the following transport mechanisms for water or moisture with dissolved aggressive substance:

- capillary suction
- permeation
- diffusion.

Cyclic wetting and drying will lead to a build-up of aggressive substance near the exposed concrete surface. The surface concentration of a substance will increase the rate at which this substance can enter deeper into the bulk of the concrete and reach the level of the reinforcement. Similarly, with one wet surface and the opposite surface exposed to drying, a one-way transport of water with dissolved substance will be created from the wet to the drying surface. This will result in an increase in the concentration of the dissolved substance, such as chlorides or sulphates, at the drying surface due to evaporation, [4].

When an aggressive substance concentrates in the outer concrete layer, it may attack the concrete or move towards the reinforcement. Here it may cause corrosion followed by cracking and spalling of the concrete cover, leading to rapid, local strength reductions.

All the transport mechanisms are non-linear by nature, except permeation, when a steady-state transport has been reached. This must be considered when the consequences of a given aggressive environment acting on a structure are evaluated. For example, the penetration depth of a carbonation front into concrete is nearly proportional to the square root of the exposure time. Chloride and sulphate diffusion will have a similar non-linear rate of penetration. The consequences of this observation are important when the optimal concrete cover is selected, [4,5].

4.3 Cracks

Most modelling of transport mechanisms assumes concrete to be homogeneous. Unfortunately, this is not the case, due to local variations in compaction, curing and, in particular, cracking.

Cracking is an often occurring feature, and load induced cracking is a natural feature in concrete sections loaded to tension. Reinforcement is introduced to compensate for this well known fact. Expansive forces due to ongoing deterioration may also lead to cracking of the concrete. This cracking may be internal or may reach the surface. It may appear as single cracks or random map cracking. In all cases, this will have considerable influence on the transport of substance into and within the concrete.

Cracks will open up the surface to early ingress of aggressive substances, including water, and the initiation period may be considerably reduced.

An example is when chloride-contaminated water enters the concrete and causes early depassivation of the reinforcement. This must be considered when initiation periods in environments containing chloride are evaluated. The rate of corrosion will then be determined by the moisture level and the availability of oxygen, parameters which, in practice, can only be roughly estimated in advance.

For carbonation, local depassivation of steel will occur earlier at cracks and corrosion may begin. However, experience shows that such purely carbonation-initiated corrosion at narrow cracks (say less than 0.3 mm at the surface) will usually die out after some time due to self-healing (autogenous healing) and repassivation of the steel, caused by clogging of the cracks with dust, rust and lime.

The transport of substance through cracks and the influence of autogenous healing is a complicated chemical and physical process. Although the autogenous healing is part of civil engineering slang, factual information about the process itself and the influencing parameters are limited. A quantification of the effect of water flow through cracks and the flow reduction due to autogenous healing was first investigated on a larger scale through the trend-setting theoretical and experimental investigations presented in [6].

The influence of micro and macro cracks on the ability of the concrete cover to provide a good protective "skin" for the structure, and to protect the reinforcement against local ingress of aggressive substances, needs much more attention than it has been given. The reliability of a service life design will depend strongly on how cracks in the cover are modelled or treated in the evaluation.

Nearly all deterioration mechanisms develop with time through two different phases, [7]:

- initiation phase: during which no noticeable weakening of the material or of the function of the structure occur, but some protective barriers are overcome by carbonation, chloride penetration or sulphate accumulation.
- propagation phase: during which active deterioration normally proceeds rapidly and in a number of cases at accelerating pace. Reinforcement corrosion is one important example of propagating deterioration.

Steel in concrete is effectively protected by the electro-chemical passivation caused by the alkalinity of the surrounding concrete. Reinforcement corrosion only occurs if depassivation has taken place. Carbonation of the concrete, ingress of chlorides, and leaching of lime can cause such depassivation. Hence these mechanisms may constitute an initiation phase, and the subsequent corrosion followed by cracking and spalling, will constitute the propagation phase.

5 Service life design

The concept of durability can be difficult to quantify and to use in practical design. This has led to the introduction of service life as the more operational term describing the intentions to ensure long term durable structures.

The CEB-FIP Model Code 1990 (MC 90), [8], expresses the basic requirements for service life design as follows:

"Concrete structures shall be designed, constructed and operated in such a way that, under the expected environmental influences, they maintain their safety, serviceability and acceptable appearance during an explicit or implicit period of time without requiring unforeseen high costs for maintenance and repair."

In practice, there are three different types of service life relevant for structures, depending on the type of performance considered:

- Technical service life: the time in service until a defined unacceptable state of deterioration is reached.
- Functional service life: the time in service until the structure becomes obsolete, from functional performance point of view, due to changed requirements.
- Economic service life: the time in service until replacement of the structure is economically more attractive than keeping the structure in service.

The functional and economic service life reflects the more common reality that structures are replaced or upgraded when they are considered obsolete for other than purely technical reasons. Usually, no prior engineering efforts can forecast such situations. The engineering input in the design must concentrate on providing technically sound and durable structures which are able, within reason, to outlive the expected functional and economic service life. The latter is based on mere educated guesswork. This guesswork can be made easier by using past experience, such as the rate at which axle loads on roads and bridges have been grown, design loads in dwellings have been increased, clearances of bridges have been raised and accidental design loads have been changed.

In both the initiation phase and the propagation phase, all significant deterioration mechanisms depend on some substance penetrating from the outside into the bulk of the concrete through the surface by one or more of the transport mechanisms described in Section 4.2.

5.1   Influence of execution

Inhomogeneity, including cracking, is one of the specific features of concrete as a structural material. The quality obtained in practice also differs from structure to structure, or from project to project, due to the dominating influence on quality caused by the execution process and by the local aggregates used for the concrete mix. These features characterise concrete as different from all other structural materials.

For existing structures, the residual service life can be predicted based on an assessment of the actual state of damage and the estimated progress of deterioration, taking the above features into account.

The penetrability of concrete cover is the combined effect of the transport mechanisms and the inhomogeneity. From this it follows that penetrability becomes one of the primary parameters to be "designed" when concrete structures are tailored to comply with specific durability requirements, such as service life. The quality obtained in the actual structure will be determined by the quality of execution and curing, and much effort must be put into ensuring this quality throughout the structure if durability is to be achieved. This is also the part of the execution which is most in need of supervision.

5.2   Quality assurance, quality control

ISO 9001 specifies quality assurance procedures in an attempt to cope with these types of problems. Unfortunately, this internationally recognised standard is aimed mainly at the factory-based production industry, and cannot reasonably be introduced in the design and construction industry without extensive modifications and at considerable cost, the warranty of which is often questionable. Due to the status of the ISO 9000 systems - and the misuse of the system by many owners and authorities - such necessary modifications are difficult to introduce.

Nevertheless, quality assurance systems compatible with the nature of the design and construction industry - and specifically coping with the inherent nature of concrete and the creation of concrete structures - are indispensable if reliability is to be maintained for this versatile building material.

5.3   Influence of structural form

The geometric form of exposed structures has considerable influence on the interaction between the concrete and the environment. Complexity in the structural form will usually increase the sensitivity of the structure to deterioration, shorten service life or require increased efforts in future maintenance. Configurations which lead to difficult execution, such as congested reinforcement, small dimensions and difficult access, increase the risks of inferior in situ quality. Such situations may eliminate all the good intentions of specifying high quality concrete, adequate covers, etc. at the design stage.

Close to out-going edges and corners, aggressive substance can penetrate into the concrete from more than one side and lead to local concentrations. If the concrete or the reinforcement is prone to deterioration under the prevailing environment, corners and edges will lead to an early development of damage at the out-going corners and along the edges - in other words a so-called corner effect develops.

6 Prediction of service life

Knowing the mechanisms of deterioration, their governing parameters, and the kinetics of the deterioration mechanisms, the parameters necessary to quantify the prediction of service life can be listed, [4].

For existing structures, the governing mechanisms of deterioration and transport have to be determined. In principle, this can be carried out based on laboratory studies of similar materials under similar exposure conditions.

However, the actual in situ value of the parameters cannot be assessed based on laboratory studies alone. A reliable service life estimate, therefore, requires a subsequent in situ verification of the values used in the initial service life design, i.e. the actual on site coefficients of transport achieved must be determined.

The primary task for long service life design is to ensure a sufficiently long initiation period. In practice, this is achieved by providing barriers against the penetration and accumulation of the aggressive substance considered. Therefore, the objective is to provide a good protective outer layer of the structures which can be carried out by the following means:

- selecting concrete quality, that is, for example, a concrete mix providing low penetrability and high chemical resistance, either by ensuring a low water-cement ratio or the addition of pozzolanic additives or both.
- selecting large concrete cover to the reinforcement, finding an optimal balance between the advantage of a larger cover and the increased risks of cracking.
- ensuring execution procedures which enhance quality, specifically in the outer layers, such as good compaction of the concrete. This requires structural dimensions and detailing of reinforcement, leaving adequate space for placing concrete and introducing vibrators. Good curing of the hardening concrete, requiring moisture control and limitation of temperature differences due to heat of hydration.

Furthermore, if they start propagating, every effort should have been made to have a reduced rate of the critical deterioration mechanisms. An example could be the use of a concrete mix with high electrical resistance. This would result in low rates of corrosion.

7 Basic protection strategies

In a previous three-man workshop, comprising Prof. Schießl from Aachen, Prof. Beeby from Leeds, and the writer, two basic protection strategies were identified which benefited from the above observations, [9]:

Strategy A: avoid the degradation reaction altogether.

Strategy B: select optimal material composition and structural detailing to resist the degradation for as long as required.

Strategy A can be subdivided into three:

A1:   change the environment by such measures as tanking, membranes, coatings

A2:   select non-reactive materials, such as stainless steel, coated steel, non-reactive aggregates, sulphate resistant low alkali cements

A3:   inhibit the reaction, by employing techniques such as cathodic protection. The avoidance of frost attack by air entrainment is also classified in this category.

Strategy B has different types of interventions. For example, corrosion protection can be achieved by selecting appropriate cover and concrete mix. However, the structure can be made more robust against aggressive environment of different sorts by appropriate detailing, such as minimizing the exposed concrete surface, introducing rounded corners or adequate drainage.

8   Design strategy for the next century

A pragmatic strategy has to be found for the service life design of new structures based on the selection of protective measures against premature deterioration.

Modelling of deterioration processes is only applicable for strategy B. The ideal procedure would start with the definition of the performance criteria related to the expected environmental conditions. The next important element is the realistic modelling of the environmental actions and the materials' resistance against these actions. Based on the performance criteria, performance tests are indispensable for quality control purposes. The performance tests must be suitable for checking the potential quality of the material under laboratory conditions and, even more important, the in situ quality.

A set of appropriate measures can be combined to ensure that the required service life is obtained with a sufficiently high probability.

This is the design concept followed by the MC90, [8], and called the "Multi-Stage Protection Strategy", to which this writer has contributed extensively, and from which parts are quoted below. It may also be considered a "Multi-Barrier Approach". This strategy leaves the selection of the individual protective measures to the designer. The different protective measures may act simultaneously, or one measure may be substituted by the next once the former has been overcome, eliminated or surpassed by the aggressive substance.

According to [8] protective measures may be established by:

- the selected structural form
- the concrete composition, including special additions or admixtures
- the reinforcement detailing including concrete cover
- a special skin concrete quality, including skin reinforcement

- limiting or preventing crack development and limiting crack widths by measures such as prestressing
- additional protective measures such as tanking, membranes or coatings, including coating of reinforcement
- specified inspection and maintenance procedures during in-service operation of the structure, including monitoring procedures
- special active protective measures, such as cathodic protection or sensor monitoring.

A different level of reliability is associated with the protective effect of each type of measure. This level depends very much upon the quality assurance scheme associated with establishing and maintaining each protective measure, [9,10].

The main problem with this design procedure is that the given material resistance parameters and the micro-environmental conditions, which show tremendous variations, are unknown at the design stage. Additionally, experience from durability failures shows that most of the failures are due to so-called gross errors, not covered by normal variations of the material properties. Therefore, it is virtually impossible, at the design stage, to quantify the characteristic material parameters that can be expected in the structure under the given environment. Neither is the situation helped by the use of more sophisticated formulae in the calculation of factors such as concrete cover required.

As a consequence of all these considerations, the use of specified material properties or properties determined from separate test specimens, cannot be used for realistic service life predictions, but only for testing of the compliance of the finished structure and its materials. The testing will determine if an acceptable service life can be realistically achieved, or whether additional measures such as surface coatings, cathodic protection or other preventive or corrective measures are needed in the future.

As a further consequence, the requirements of codes ensuring sufficient durability need to be given in a prescriptive form.

Thus, requirements aiming at the concrete mix (cement type and content, w/c-ratio, etc.), design parameters (cover, detailing etc.), and execution (compaction, curing, etc.), while being essential to determine the accept/reject criteria for the product, will not serve as a basis for realistic service life calculations.

From the above outlined problems, it can be concluded that calculation of true service life at the design stage is an unrealistic procedure, whereas calculations predicting future performance will be very valid and useful when assessing existing structures, providing adequate test methods exist to define the actual material properties in the structure. Much more work is needed in the development of in situ test methods, but it is already possible to carry out useful predictions in many cases. For example, if the carbonation depths are measured in a structure that is 5 years old, the time when carbonation will reach the reinforcement can be predicted with reasonable reliability, and any necessary remedial procedures can be planned, [11].

Accessories such as drainage, joints, bearings, railings, connections, installations, fixtures etc. usually have a shorter service life than the structure itself, and adequate provisions for maintenance and replacement should be provided for in the design.

It must be emphasised that design for a specific service life does not mean that the structure will perform satisfactorily during the whole service life without maintenance and repair. On the contrary, some degree of inspection and maintenance is considered an integral part of the service life design.

8.1 Additional protective measures

Concrete is an excellent and durable building material, also under aggressive hot and humid environments, provided it is used correctly. With few exceptions, the durability of structures and long satisfactory service life - in excess of say 50 years - can be achieved using good quality concrete in a well-designed structure, without additional protective measures.

The few, though important, exceptions where additional protection may be needed relate to extreme exposure conditions. Such situations may be:

- foundations and other buried structures in moist soils heavily contaminated by sulphates and chlorides
- marine splash zones and other similar areas where cyclic wetting and drying with similarly contaminated water continues to "pump" the aggressive substance into the concrete, e.g. drydocks, marine jetties, harbours and cooling canals
- special industrial plants with liquid or gaseous aggressive chemicals, such as refineries, petrochemical plants and desalination plants.

The choice of protective measures must be carefully considered in relation to the particular aggressive environment encountered. Possible secondary effects must also be considered, such as the selection of epoxy-coated reinforcement with individually coated bars, which will rule out the later use of cathodic protection. Another unfavourable secondary effect could be the increased rate of carbonation following treatment with water repellant impregnations.

Additional protection is considered to be measures such as, [8]:

- special additives enhancing the impermeability of concrete
- special admixtures neutralising or inhibiting deterioration mechanisms
- coating to either steel or concrete
- bituminous or polymeric protective membranes
- physical linings, such as sacrificial steel linings (piles), stainless steel linings
- electrochemical protection (CP)
- non-corrodible reinforcement such as stainless steel, polymer fibre bars
- fibre reinforcement (polypropylene, glass, steel).

Before additional measures of the type listed above are selected, the concrete mix should be considered and the type of binder chosen.

Portland cement is the traditional binder and variations in composition are mainly dictated by the presence or absence of sulphates - ASTM Type V cement is used to protect the structure against sulphate attack. This type of sulphate resistant cement has limited capacity to bind chlorides and thus a lower threshold value for chloride corrosion.

During the past decade, pozzolanic additions have become more and more used due to their beneficial effects in enhancing durability. Such additions have mostly been microsilica, flyash or slag, all available in abundance, and all able to neutralise the effects of alkali-silica reactions. To some extent, therefore, they compensate for alkali reactive aggregates and may offer very low diffusivity for chloride and sulphate ions.

8.2   Reliability of new protective products

Recent years have seen the development and marketing of numerous products claiming to "solve" the durability problem in design, or "elegantly" to prevent further deterioration of existing structures.

Due to the severe economic consequences of introducing costly protective measures, which in the end turn out not to perform as expected (and in some case may even have adverse effects), true site performance of such additional protective measures must be documented to a reasonable extent. It is evident that very long-term testing and documentation is not realistic for all these product due simply to the long time scales involved. In fact, this is one of the main problems facing the construction industry. The questions that arise time and time again are:

- how can organic materials provide long term protection of inorganic materials such as steel and concrete?
- can we have faith in chemically active additives that have to be dormant in the concrete structure for many years before their active protection is required? Do they not diffuse or wash out of the concrete in the meantime? To put it bluntly, are we just paying for hot air?
- why has epoxy coating of steel pipes in aggressive soils performed satisfactorily for many decades, and yet epoxy coated reinforcement seems to cause problems after 10 to 15 years? Has the industry used the right technology? Has the industry used the right formulations? Has the coating industry told us the whole story?

Many such question are asked, and the more examples of non-performance reported, the more difficult it becomes for our concrete construction industry to maintain credibility.

Due to the inherent time scale problem, the only way to compensate for lack of long-term experience with new products, is for clear scientific models to be established, capable of describing the deterioration mechanisms the material and the structure can be exposed to, and to model how the new material will influence these mechanisms. Accelerated testing, combined with true in situ testing, can then verify, and pinpoint the adjustment or, in some cases, rejection of the material. A controversy exists today between the time scale of the construction industry and the time scale of the mass production industries - the former operates in decades, the latter over a few short years.

It is disappointing to experience that producers consider 2 or 3 years experience with a product, or a whole repair concept, to be "valuable long term documentation", when we know that nothing can be documented under real environmental exposure within that time scale. An example - repairing chloride-induced corrosion can be done either by traditional patch repairs or by one of the electro-chemical repair methods. To present experience from one, two or even three years of performance of the repaired structure as "documentation" of good performance is absurd. In most cases, no matter how bad the patch repair was made, it would still look "good" after that short period of time. Most of our deterioration mechanisms develop much slower than that.

In this respect, the Middle East, in particular the Gulf, is a very special region, because of the degree of aggressiveness of the environment and the high temperatures which greatly accelerate deterioration processes under natural conditions, [2]. In many other parts of the world such "accelerated in situ testing" would be considered a very valuable form of accelerated testing. The writer has, on several occasions, encouraged the repair industry and the repair product industry to take more advantage of this type of realistic accelerated testing to demonstrate the value of their products. It is disappointing to see how few have actually taken up this challenge; perhaps they do not have faith in their own products? However, some very valuable tests of this sort have been made. In particular reference shall be made to the long term testing of carbonation reducing surface coatings initiated by the former Ministry of Works, Power and Water in Bahrain, in cooperation with BRE in the UK, to make comparative testing on identical specimens treated identically and then exposed to both UK and Bahrain environments. Results from these tests have been reported systematically, and should stand as a model for others to follow.

9   The built environment of the 21st century

To be able to select the most reliable protective measures and to achieve longevity of our structures, a careful strategy should be followed through all phases of the creation of structures. This involves the designer, the contractor and producer, the owner and the user.

- The designer should base his design on a clearly formulated design strategy, where the true environment is identified and the required performance and target service life is clarified and agreed with the Owner/Client. The specification developed for the individual structure should reflect the intentions of the design strategy followed, should specify the required performance criteria, the compliance tests to be used, the acceptance criteria and, as an important addition often forgotten or ignored, should specify procedures to be followed in the cases of non-conformance with the requirements. This latter issue is much in need of practical development. The tendency today is to consider requirements of the specs as on/off requirements. This situation has created one of the most serious drawbacks in contract handling within the construction industry.

  No one benefits from the controversies arising from the present on/off approach,- except the lawyers. Maybe they are the pariahs in this whole process. Take the situation in the US, for example, regarding liability and its consequences as a frightening example. The amount of energy, time and money being used in tackling legal aspects of design and construction contracts is deploring. This energy should be diverted into sound engineering efforts, concentrated on finding the optimal solutions to the problems or tasks to be solved. This would be to the great benefit for the Owner/Client as well as to society in general.

  At the present time, the development within the construction industry is heading in the wrong direction. There is only one source from where a true change in this deplorable trend may be changed for the better - the owner, the person that has to pay in the end. Not before the owner, in particular public owners and society, has realised the dramatic consequences of the present development, can improvements be expected. Another aspect of this quality problem is associated with the low-bid syndrome discussed separately.

* The contractor should execute the structures according to the design and the specs. This is obvious - but what is becoming more and more important is that the contractor himself should think, meaning he should make an effort to understand the project and the intentions designed into the project. The reason for this need is the same as the reasons why many codes of practice start by requiring that the follower of the code should have sufficient engineering competence and experience to understand the code. A similar requirement, stating that the personnel entrusted with this project should document sufficient professional competence and experience to understand it and the intentions of the specs, would be a valuable addition in normal construction contracts

    The reason for such a complete need for project understanding is based on the fact that the specifications cannot realistically contain instructions on every single detail needed to achieve the final goal. In practice, this would be overwhelming and confusing in practice.

* The owner should appreciate that he is an integral part of achieving the correct final solution. He may, by his actions or lack of actions, either facilitate the whole process or make it extremely complicated by excessively formalising the relationship between the partners involved, that is the designer, contractor and owner.

* The user also plays an important role by ensuring that there is a follow-up on the structure and its performance during the period of use. Often the Client is also the Owner. The acceptance that, in many respects, concrete structures are living structures, with properties changing during use, and with ageing characteristics, has apparently slipped the minds of many owners. This is surprising, as all owners are aware that steel structures need regular repainting, and timber structures need regular treatment with fungicide. Concrete structures, too, need some form of regular maintenance in order to ensure long service life. The interventions needed depend, of course, on the type and aggressivity of the environment in which the structure is performing.

    For large civil structures or large plants or installations, a regular inspection and maintenance programme is needed to carry out the correct maintenance of structures to ensure their reliable performance. When designing such structures, it is important to know whether such maintenance procedures can be relied upon. The conditions under which the structure is to be used are an integral part of the initial design procedures by determining which strategy to follow when designing for a specific target service life.

When considering the problems facing the construction industry today, it is valuable to realise that the problems have been clearly identified and are actually simple and easily understood. The possible solutions have also been identified and, to a large extent, even quantified - they, too, are simple. What remains is to tackle the prime remaining problem - how to ensure that Owners/Clients understand the nature of the very specific problems associated with creating durable and well-performing concrete structures - and that they are aware of the consequences to finance and quality that follow in the wake of their decisions.

The role of the client and the role of the legislation in either promoting or adversely influencing the build environment is a subject much in need of discussion and clarification. During the past decade, developments have been adverse, governed more and more by non-technical issues, towards a dominating influence of the legal system in the form of shrugging off responsibility. Responsibility can only be delegated by taking out insurance which does not, in itself, provide any guarantee of quality. It does, however, clarify the economic conditions. What is important is to realise that the client cannot delegate his responsibility to designers or contractors. Only by an all-risk insurance taken out by himself, or by paying double for insurances taken out by the consultant and the contractor, can he become financially free of responsibility - which does not, by itself, enhance durability or quality either.

Consequently, improved performance of concrete structures built for the next century will depend on the willingness - and ability - of all partners in our build environment to cooperate in creating the new and advanced concrete structures designed for the 21st century. If this is achieved, a giant leap forward is predicted in the application of concrete for constructing reliable and durable structures breaking new ground in height, span, size and appearance.

10　Conclusion

Classical procedures for the design, construction and use of concrete structures have failed to provide reliable long-term performance of structures exposed to aggressive environments. During the next century, world societies will be faced with immense challenges in solving the explosion in building and infrastructural needs generated by growing populations and increasing demands for quality. Improvements of the current construction qualities are needed in service life performance and reduced maintenance costs. This will require multi-disciplinary engineering education and training in contrast to the specialisation of today. It will also require a greatly increased public recognition of the long-term technical and economic consequences of alternative solutions. The long-term costs of short-term savings will give a backlash on the engineering profession if the true decision makers are not made aware of their responsibilities. These decision makers are the owners, often guided by political decisions, and influenced by the regrettable myopia of the media, [11]. Hence, the engineering profession is faced with the challenge of improving their ability to communicate, a challenge which is just as important as their ability to solve complex technical and economic problems.

11　References

1.　CEB-RILEM International Workshop. (1983) Durability of Concrete Structures (ed. Steen Rostam), Workshop Report, *CEB Bulletin d'Information* No.152, 1984.

2. Rostam, Steen. (1991) Philosophy of Assessment and Repair of Concrete Structures, and the Feedback into New Designs. Keynote Lecture, *Regional Conference on Damage, Repair Techniques and Strategies for Reinforced Concrete, Proceedings.* Bahrain, 7-9 December. pp. 33-126.

3. CEB, Comité Euro-International du Béton. (1982) Durability of Concrete Structures, State-of-the-Art Report.*CEB Bulletin d'Information,* No. 148.

4. CEB, Comité Euro-International du Béton. (1992)*Durable Concrete Structures, Design Guide,* Thomas Telford, London.

5. Schießl, Peter. (1976) Zur Frage der Zulässigen Rissbreite und der Erforderlichen Betondeckung im Stahlbetonbau - unter Besonderer Berücksichtigung der Karbonatisierung des Betons. Thesis, *Deutscher Ausschuss für Stahlbeton,* Heft 255, Berlin.

6. Edvardsen, Carola. (1996) Wasserdurchlässigkeit und Selbstheilung von Trennrissen in Beton (Water Permeability and Autogenous Healing of Cracks in Concrete). Thesis, Institute of Building Materials Research, ibac. University of Technology, Aachen, 1994. See also *Deutscher Ausschuss für Stahlbeton,* Heft 455, Berlin.

7. Tuutti, Kyösti. (1982) Corrosion of Steel in Concrete. *Swedish Cement and Concrete Research Institute, CBI,* Fo 4:82. Stockholm.

8. CEB, Comité Euro-International du Béton. (1993) *CEB-FIP Model Code 1990, Design Code.* Thomas Telford, London.

9. Siemes, A.J.M. and Rostam, S. (1996) Durable Safety and Serviceability, a Performance Based Design. IABSE Colloquium, *Basis of Design and Actions on Structures, Proceedings.* 27-29 March, Delft.

10. Rostam, Steen and Faber, Michael. (1996) Rehabilitation Strategies - A Multidisciplinary Challenge to the Engineering Profession. International Conference, *Concrete in the Service of Mankind, Proceedings,* Volume V. 24-28 June, Dundee.

11. Rostam, Steen and Schießl, Peter. (1994) Service Life Design in Practice; Today and Tomorrow. International Conference, *Concrete Across Borders, Proceedings,* Volume I, 22-25 June, Odense.

7   IMPORTANCE OF PREDICTION TECHNIQUES ON DURABILITY IN THE DURABILITY DESIGN FOR CONCRETE STRUCTURES

S. NAKANE and H. SAITO
Obayashi Corporation, Technical Research Institute, Tokyo, Japan

Abstract
In present concrete structures design, concrete durability is not always taken into account quantitatively. This paper presents a method for realizing this concept.

First, degradation phenomena related to the durability of concrete structures are listed. Then a method of clarifying the deterioration mechanisms is described, using a Fault Tree Analysis as an example. For quantitative treatment of the durability problem in design, it is necessary to quantitatively comprehend the deterioration mechanism. This has not been satisfactorily achieved up to now, because the deterioration mechanisms are so complicated. Therefore, it is emphasized that all deterioration mechanisms affecting concrete durability have to be clarified and the data required to quantitatively predict deterioration must be accumulated.

In the latter part of this process, durability predication techniques must be developed. The authors have been conducting research works on deterioration of concrete in contact with water. From this research, an acceleration test method was established which is useful in quantitatively predicting long term deterioration of concrete. This test method is described as an example of a durability prediction technique, and its applicability is discussed.
Keyword: Acceleration test, chemical attack, deterioration phenomena, durability, leaching.

1 Introduction

It would be better to design concrete structures from the stand point of structural performance, considering time-dependent behaviors of durability. However, actual

Integrated Design and Environmental Issues in Concrete Technology. Edited by K. Sakai. Published in 1996 by E & FN Spon, 2–6 Boundary Row, London, SE1 8HN, UK. ISBN 0 419 22180 8.

designs are forced to be executed not directly linked to durability, because confidential prediction methods for durability have not been established.

Deterioration factors affecting durability are carbonation of concrete and corrosion of reinforcing bars, which give rise to concrete cracking, freezing and thawing action, chemical attack, heat action, abrasion and so on. These factors should be suitably taken into account in an actual design, based on environmental conditions and the purpose of the structure. Carbonation is the only factor that can presently be applied quantitatively in the durability design. The others are subjects for future study.

This paper presents a concept for taking into account concrete durability in design. Macroscopic degradation phenomena related to concrete structures are classified and Fault Tree Analyses are shown using some actual applications to clarify degradation mechanisms. Finally, as an example of a durability prediction technique, an acceleration test method established by the authors for concrete in nuclear waste repositories, together with an application of this method, is presented.

2 Degradation phenomena and their mechanisms

2.1 Degradation phenomena
Degradation phenomena affecting the durability of concrete structures are listed in

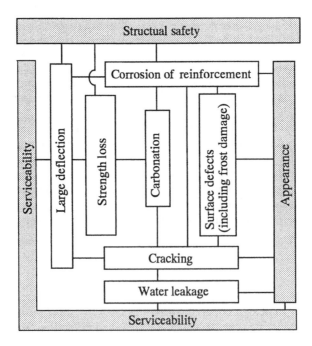

Fig. 1.   Correlative relationship between degradation phenomena
and corresponding performances of structure

Fig.1, which shows correlative relationships among them and their effects to the performances of structure. The 7 internal items in the non-hatched frame are major degradation phenomena and the surrounding hatched items indicate performances of structure. Here, frost damage and abrasion are included in "Surface defects" phenomena.

2.2 Deterioration mechanisms

Each degradation phenomenon is caused by many deterioration factors. These factors are very complicated, which has made durability design very difficult.

The authors have aimed at a Fault Tree Analysis (FTA) as a tool to clarify deterioration mechanisms. This method was originally developed in the field of reliability engineering. In this analysis an undesirable phenomenon is taken up at first. Then all sub-phenomena which are causes of the first phenomenon are picked up and connected with a logic gate of "AND" or "OR", considering the relation of cause and effect. Furthermore, factors to cause the sub-phenomena are studied and these procedures are repeated until factors can not be broken up anymore. Thus, a deterioration mechanism tree of tournament style is constructed, where the following symbols are used[1] :

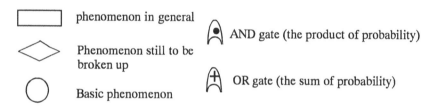

In the use for which FTA was originally developed, it is possible to make a quantitative evaluation of an undesirable phenomenon through its mechanism and the probability of each factor.

When FTA is applied to the durability problem of concrete, a quantitative treatment can not be realized at present. However, it is still useful in explaining the overall deterioration mechanism, and a qualitative evaluation will become possible if the routes of the tree can be graded.

From this point of view, Fault Tree Analyses of the degradation phenomena shown in Fig.1 were carried out and 2 examples, FTA of "Strength loss" and "Surface defects" of concrete, are shown in Fig.2 and Fig.3, respectively. In these figures, the symbol of basic phenomena defined above and enclosed by circles are used as to express the countermeasures to be taken to improve concrete durability. The thick line connecting the phenomena indicates that the route has a relatively strong relation in the FTA in the depicted case.

Through a Fault Tree Analysis, a more clear and systematic understanding of the complicated durability problem can be obtained. However, an analysis by this method tends to depend on personal subjective judgment. Therefore, many trials of FTA should be executed by many different researchers to get a consistent FTA model of the concrete durability problem. At the same time, an effort should be made to obtain more quantitative relations in FTA.

Fig. 4 shows the FTA of concrete corrosion, which is derived from phenomenon

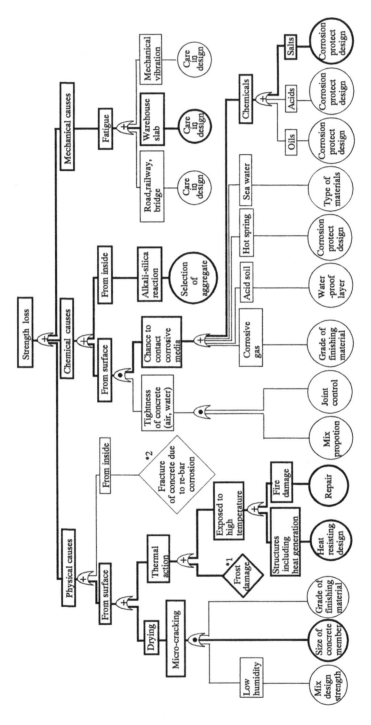

[Note] : Factors concering bad strength gain before complete hardening are not contained.
*1 and *2 shall be referred to the FTA of the other deterioration phenomena.

Fig. 2.   Fault tree of the strength loss in hardened concrete

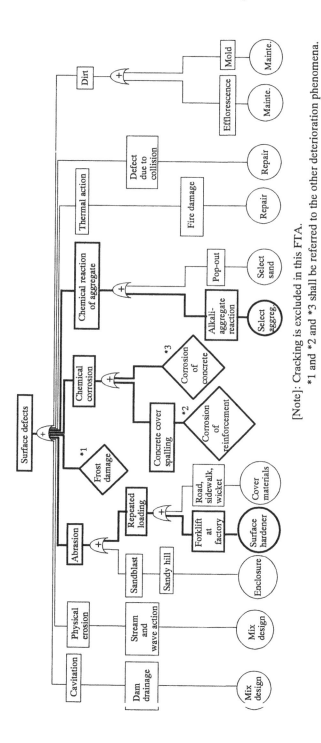

[Note]: Cracking is excluded in this FTA.
*1 and *2 and *3 shall be referred to the other deterioration phenomena.

F ig. 3.   Fault tree of surface defects of concrete

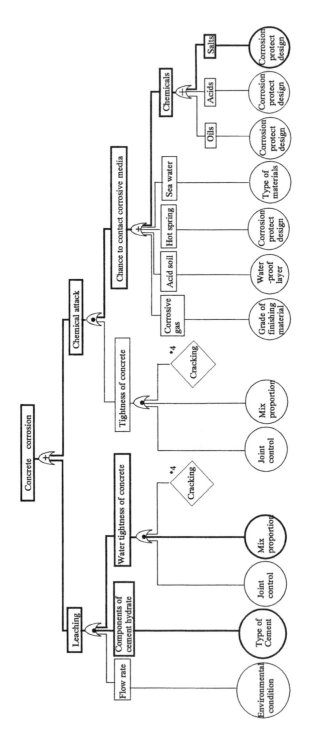

[Note] : *4 shall be referred to the FTA of cracking

Fig. 4. Fault tree of concrete corrosion

*3 in Fig. 3. Here, the main deterioration sub-phenomena are grouped into two classes based on the nature of the corrosives. Cement hydrate in contact with water dissolves gradually even where there are no corrosive media. This phenomenon is generally called leaching. Where the water contains corrosive media, cement hydrate reacts with them and produces characteristic deteriorated minerals.

3 An acceleration test method for concrete contacted with water

In the design of nuclear waste repositories where concrete may be used as a barrier and/or structural material, degradation prediction is required in the range of over 1,000 years. For this purpose, an acceleration test is a useful method for quantitatively predicting deterioration. The authors have proposed a new test method for predicting the chemical deterioration of concrete in contact with water [2]. This section outlines the test method and its applicability. The scope of the investigation closely corresponds to the thick line of FTA in Fig. 4.

3.1 Outline of the test method

The degradation mechanism of concrete in contact with water can be explained as follows. Where water is pure, cement hydrate structure will gradually change as cement hydrate gradually dissolves in water for a long time period. This phenomenon is caused by diffusion of Ca^{2+} into water, as its concentration in pore solution is higher than in water. Where water contains some corrosive media such as Cl^- or SO_4^{2-}, besides above mentioned phenomenon, corrosives filter into cement hydrate structure, react with hydrate and produce characteristic minerals. Generally, these produced minerals are more dissoluble than hydrate. In this case total deterioration speed becomes larger than that without corrosives.

Fig. 5 illustrates the experimentation apparatus, where the specimen is placed between two glass vessels, each containing one liter of water. Each vessel has an anode and cathode electrode, connected via an electric DC power source and an ammeter, to provide a potential gradient across the specimen.

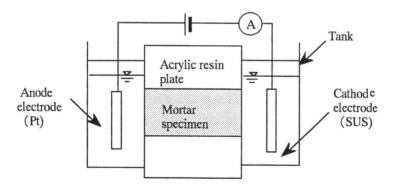

Fig. 5. Experimentation apparatus

Table 1. Specimen and test condition

Specimen	Test condition
OPC : Sand : Water = 1 : 2 : 0.65 (wt.) 50 mm in diameter, 30 mm high	Potential gradient : 0, 2, 5, 10 V/cm Solution : Ion exchanged water Temperature : Room temperacure (25 °C) Test period : 2, 5, 8, 12 months

The specimen is cylindrical and made of mortar or concrete. The shape of the specimen can be freely chosen for the particular test.

The quantity of Ca^{2+} dissolved in water is measured as the index for judging the degree of acceleration. The solution around the cathode is collected periodically, exchanged once a week, and its Ca^{2+}, Na^+ and K^+ concentrations are determined by atomic absorption analysis. Its pH level and conductivity, as well as the quantity of current passed, are monitored for reference.

The specimen is taken out after the test and cut with a diamond cutter to view its cross section. The condition of its cement hydrate structure is observed and its deteriorated thickness is measured with a rule. Then it is divided into two pieces at the deterioration boundary and the degree of deterioration of cement hydrate is measured by the following methods.

• Component of cement hydrate: X-ray diffraction method
• Amount of $Ca(OH)_2$ and $CaCO_3$: thermal analysis
• Ca/Si mole ratio of C-S-H: chemical analysis
• Pore distribution: mercury injection method

3.2 Application possibility to leaching deterioration

3.2.1 Effect of potential gradient on dissolving speed of Ca^{2+}

A test was carried out under conditions shown in Table 1,   to clarify the effect of potential gradient on dissolving speed of $Ca^{2+}$and deteriorating speed of a specimen. Four potential gradients, 0, 2, 5 and 10 V/cm, were applied.

Fig. 6 shows the cumulative quantity of Ca^{2+} dissolved in water. Ca^{2+} increased gradually up to 388 mg at the 365th day under zero potential gradient.

Applying a potential gradient to the specimen caused significantly accelerated dissolution of Ca^{2+} with a cumulative quantity reaching 3,658 mg at the 365th day under 2 V/cm, 6,104 mg under 5 V/cm and 8,337 mg under 10 V/cm.

Fig. 7 shows the relationship between the cumulative quantity of Ca^{2+} and the potential gradient. The amount of Ca^{2+} dissolved in water was almost proportional to the potential gradient within the range of 2 to 10 V/cm. The extent of acceleration with increase in potential gradient was estimated by comparing the Ca^{2+} dissolution rates.   It was calculated by comparing number of days required to produce the same quantities of Ca^{2+} under different potential gradients, using data over 12 months. The acceleration rate was 13 times greater under 2 V/cm, 30 times under 5 V/cm and 60 times under 10 V/cm, compared with that under the zero gradient condition.

Fig. 8 shows the relationship between the cumulative quantity of Ca^{2+} and the deteriorated thickness. The deteriorated thickness was clearly observed and it

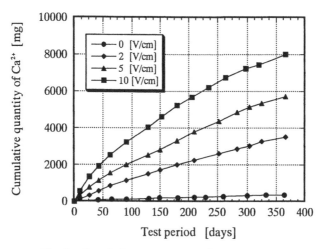

Fig. 6.   Cumulative quantity of Ca²⁺ dissolved in water

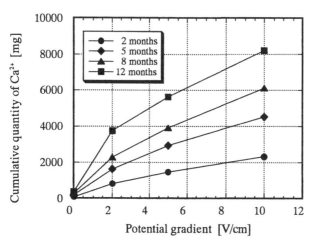

Fig. 7.   Relationship between cumurative quantity
of Ca²⁺ and potential gradient

corresponds to the cumulative quantity of Ca^{2+} under each potential gradient. This shows that deteriorating speed is also similar to dissolving speed.

3.2.2   Deterioration of cement hydrate structure

It is presumed that the deterioration of a cement hydrate structure does not occur uniformly throughout the concrete, but progresses gradually from surface to interior over a long period. The test was continued for 12 months under 5 V/cm potential

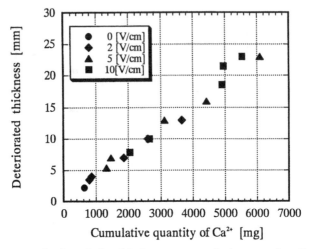

Fig. 8. Relationship between cumulative quantity of
Ca²⁺ and deteriorated thickness

Cathodic side	2mm	20 mm	10 mm	
Region	I	II	III	IV
Components of hydrate	SiO₂	C-S-H	C-S-H (CH)	CH, C-S-H
Ca/Si mole ratio of C-S-H	(0)	(1.2)	(1.8)	(1.9)
Total pore volume	-	0.158	0.113	0.109

Fig. 9.   Deterioration of the specimen

gradient, using specimens of the same mix proportions described in 3.2.1 except size (20 mm in diameter and 100 mm high).

Fig. 9 illustrates the deterioration of the specimen. Two boundaries were clearly distinguished by visual inspection. One was about 2 mm from the side facing the cathode. This 2 mm-thick region was brown and very porous. The other boundary was about 30 mm from the first one. This 30 mm-thick region was coarser than the non-deteriorated inside region, which was dense. Fig. 9 also illustrates the deterioration of the specimen, determined by changes in the components of cement hydrate obtained by X-ray diffraction, changes in the Ca/Si mole ratio of C-S-H determined by chemical analysis and changes in total pore volume measured by mercury injection. The deteriorated part was classified into four regions (I ∼ IV), corresponding to the degree of deterioration. The boundary between II and III is provisional, because it was decided for the convenience of sampling and the values obtained were average in that region. Each region can be characterized as follows.

Region I : the surface facing the cathode ~2 mm
C-S-H was completely converted to silica gel and the cement hydrate structure was very porous.
Region II : 2 mm~22 mm
Almost of $Ca(OH)_2$ dissolved, C-S-H was slightly deteriorated and cement hydrate structure was slightly porous.
Region III : 22~32 mm
A little $Ca(OH)_2$ remained, C-S-H was not deteriorated and cement hydrate structure was very slightly porous.
Region IV : 32 mm~
The cement hydrates were not deteriorated.

These results show that the deterioration of cement hydrate progresses gradually from surface to interior. This test method can reproduce the actual deterioration mechanism, that is, first, dissolution of $Ca(OH)_2$, then gradual dissolution of C-S-H [3]. It can be said that this test method is quite useful in simulating cement hydrate deterioration by leaching.

3.3 Application possibility to deterioration by chemical attack

Deep underground water may contain chemical components that are harmful to concrete. However, there is no test method established to evaluate their effects. Thus, the acceleration test was executed under a 10 V/cm potential gradient condition, using the same specimen as described in 3.2.2. Solutions used in cathodic side cell were 300 mg/l Cl^-, 100 mg/l SO_4^{2-}, 15 mg/l HCO_3^- and a mixture of these three solutions. These amounts were decided from mesured examples of chemical components in deep ground water. Ion exchanged water was also used for reference.

Fig.10 shows the cumulative quantity of Ca^{2+} dissolved in each solution. The amount of dissolved Ca^{2+} was different in each solution. Furthermore, compared to the ion exchanged water, there was 45% more in chloride ion, 35% less in sulfide ion, and 80% less in carbonate ion. The very small amount of Ca^{2+} dissolved in carbonate ion might be due to sedimentation of $CaCO_3$ on the concrete surface at the cathode. These data show that there are different amounts of Ca^{2+} in the different solutions.

Fig.11 shows photographs of the specimens after tests. It shows the cut portion of the specimens of the same size to emphasize deteriorated region. Results differed significantly depending on the condition whether it contained sulfate ion or not. When sulfate ion was not contained, deterioration could be clearly distinguished by visual inspection of the cathodic regions. The non-deteriorated region was dense in texture and dark gray, while the cathodic region was coarser and light gray. The thickness of the deteriorated regions were different among the solutions. It was 18 mm in the ion exchanged water, 24 mm in the water containing chloride ions and 8 mm in the water containing carbonate ions. On the other hand, when sulfate ions were contained, there was evidence of expansion in the specimen. The thickness of the deteriorated region was about 40 mm in the water containing sulfate ions only, and about 45 mm in the mixed solution.

Table 2 shows the results of hydrates investigated by X-ray diffraction. There were $Ca(OH)_2$ and C-S-H in the regions that were determined as non-deteriorated by visual inspection. Friedel-salts were also present in the case of water containing chloride ions. However, only C-S-H was present in the severely deteriorated regions.

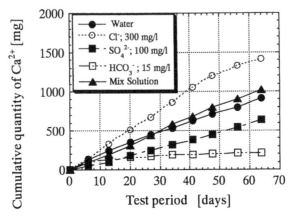

Fig. 10. Cumulative quantity of Ca^{2+} dissolved in water

Cathodic side

Table 2. Cement hydrate constitution by X-ray diffraction

	0 20 40 60 80 100 mm	Remarks
	Cathodic side Anodic side	
NON	C-S-H CH, C-S-H	NON : Ion exchenged water
CL	C-S-H CH, C-S-H, CH, C-S-H F-salt	CL : Chloride ion 300 mg/l SO : Sulfate ion 100 mg/l HC : Carbonate ion 15 mg/l
SO	C-S-H CH, C-S-H	MIX : Mix solution C-S-H : $xCaO \cdot ySiO_2 \cdot zH_2O$
HC	C-S-H, CC CH, C-S-H	CH : $Ca(OH)_2$
MIX	C-S-H CH, C-S-H, CH, C-S-H F-salt	CC : $CaCO_3$ F-salt : Friedel salt

$CaCO_3$ was also detected in the case of water containing carbonate ions. However, ettringite, which is assumed to cause expansion, was not detected. In all cases, the actual deterioration characteristics were reproduced in this test, except the presence of ettringite. The reason for the non-detection of ettringite seems that, because of its high solubility, it dissolved as soon as it formed. Furthermore, since there was no $Ca(OH)_2$ in the deteriorated region, it is clear that the dissolution of cement hydrate also occured in this test.

The above mentioned results show that there are differences among the quantities of Ca^{2+} dissolved in each solution due to cement hydrate dissolution, and this method can reproduce the characteristics of deterioration by chemical attack. Therefore, it is concluded that the method has the possibility to be applied as the accelerated test of chemical attack by dilute corrosive solutions.

3.4 Future prospects of the test method
It has become clear that the acceleration test can reproduce the actual deterioration process of concrete in contact with water and has a possibility to predict such deterioration phenomena within a short time. The method will be refined hereafter to establish an evaluation system for converting accelerated test data into deterioration in real time.

4 Concluding remarks

At present in the design of concrete structures, durability of concrete is not always taken into account quantitatively. This may be because degradation phenomena are very complicated and few data are available for quantitative prediction. This paper presented a concept for taking into account concrete durability in design. Main conclusions are as follows :
    (1) Fault Tree Analysis is a very useful method for understanding complicated phenomena of durability in a systematic way.

(2) The acceleration test method is introduced as a sample of quantitative prediction technique on durability of concrete in contact with water. The method is quite useful in simulating cement hydrate deterioration by leaching and has the possibility to be applied as an accelerated test of chemical attack by dilute corrosive solutions.

5 References

1. Barlow, R. E. and Poroschan, F. (1975) *Statistical Theory of Reliability and Life Testing,* Holt, Rinehart and Winston Inc.
2. Saito, H., Nakane, S. and Fujiwara, A. (1992) Preliminary experimental study on the deterioration of cementitious materials by acceleration method, *Nuclear Engineering and Design,* Vol. 138, No. 2, pp.151-155.
3. Brown, P. W. and Clifton, J. R. (1988) Mechanisms of deterioration in cement-based materials and in lime mortar, *Durability of Building Materials,* 5, pp. 409-420.

8 COUPLED MASS TRANSPORT, HYDRATION AND STRUCTURE FORMATION THEORY FOR DURABILITY DESIGN OF CONCRETE STRUCTURES

K. MAEKAWA, R.P CHAUBE and T. KISHI
The University of Tokyo, Tokyo, Japan

Abstract
Strength and microstructure development of cementitious materials are crucial factors that control the various parameters defining the durability of concrete. This paper seeks to establish a rational computational framework for the durability analysis of concrete structures by quantitatively evaluating such factors. This is done by studying the early age development of concrete with regard to microstructure and strength development. The method involves dynamic coupling of cement hydration, moisture transport and microstructure formation models. In this way development of pore structure at a point can be traced with increase in the degree of hydration for arbitrary temperature and moisture content history. As a verification target, the effect of different curing conditions on the strength gain, moisture loss and the pore structure development for specimens of different mix proportions and curing condition is quantitatively investigated.
Keywords: Durability, cementitious materials, microstructure development, hydration, moisture transport, curing.

1 Introduction

There exists a huge body of knowledge on how to make structural concrete durable in terms of the material, construction and structural detailing of the reinforcement. In fact, codes of practice have described how material, construction and reinforcement etc. shall be managed. Here, it must be pointed out that the statement had been qualitative and plenty of codes used to be like "engineering notes" in comparison with structural design codes for safety requirements. For quantitatively systematizing the engineering knowledge, a durability design proposal[1] was first issued by JSCE in 1989 and internationally introduced at the occasion of the CEB plenary session in 1989. The original and crucial point of this proposal concerning durability engineering

Integrated Design and Environmental Issues in Concrete Technology. Edited by K. Sakai. Published in 1996 by E & FN Spon, 2–6 Boundary Row, London, SE1 8HN, UK. ISBN 0 419 22180 8.

was that it numerically scored the overall durability performance by using some durability points, with which the durability of concrete structures can be objectively ranked. Here, the new concept made clear that the evaluation of the limit state of durability of concrete structures is the core of development.

It has to be noted, however, that the scoring procedure of durability at this moment is in a primitive stage with simplified empirical formulas, because a computational approach which enables us to predict aging and deterioration of structural concrete in space and time domains is still under investigation. Further research and development along the line of verifying durability is required to make durability design sophisticated and more versatile. With this background, the authors propose a new scheme of checking the durability limit state using a full computational approach based on coupled mass transport, hydration and micro pore structure formation theory.

Here, it will be meaningful to overview structural design with regard to the safety and serviceability limit states. Owing to the development of enhanced constitutive laws for concrete and reinforced concrete [12], it is now possible to numerically predict the structural response and the mechanical states of constituent materials in time and 3D space under any external mechanical action, even though further research is still needed. This generic structural analysis method which derives from microscopic modeling of concrete can be used for examining macroscopic structural performance with respect to the limit states to be examined. The objective of this paper is to seek for so called life-span simulator of structural concrete based on the microscopic modeling for concrete corresponding to the generic structural analysis method which is being established in structural engineering.

2 Examination of durability performance

The JSCE durability design proposal [1] is summarized in Fig.1. The overall durability of structural concrete is numerically scored according to the following linear summation rule.

$$DI = F(\phi_1, \phi_2, \cdots, \phi_n) = \sum_{k=1}^{n} F_k(\phi_k) \tag{1}$$

where ϕ_k is the influencing factor on the durability performance such as water to cement ratio, water content, slump and spacing of reinforcement, etc, and DI is defined as durability index of scalar value. In fact, the influencing factors interact nonlinearly but the linear summation hypothesis is accepted as the first assumption. On quantitatively and comparatively defining the weight of importance F_k of each influencing factor on DI, empirical formulas were primarily used since the theoretical approach was never established.

The above equation is used for only checking the durability limit state, that is, "50 years maintenance free" condition as,

$$DI \geq DI_{req} \qquad DI_{req} = 100 \quad \textit{for usual environment} \tag{2}$$

where DI_{req} is the required value of index, which is changed in accordance with the environmental conditions.

This approach enables us to easily compute the durability index, and to actualize durability performance of reinforced concrete. However, when the durability limit state is changed, the function F and F_k in Eq.(1) and the requirement in Eq.(2) have to be modified for each limit state. Since the versatility is limited, the durability examination method in Eq.(1) and Eq.(2) can be regarded as a tentative method. Here, it is crucial at present for us to get a clear idea of the idealized method of evaluating durability performance of structural concrete in future durability design.

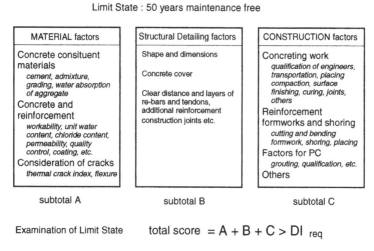

Scoring Scheme of Durability Performance
Limit State : 50 years maintenance free

MATERIAL factors	Structural Detailing factors	CONSTRUCTION factors
Concrete consituent materials *cement, admixture, grading, water absorption of aggregate* Concrete and reinforcement *workability, unit water content, chloride content, permeability, quality control, coating, etc.* Consideration of cracks *thermal crack index, flexure*	Shape and dimensions Concrete cover Clear distance and layers of re-bars and tendons, additional reinforcement construction joints etc.	Concreting work *qualification of engineers, transportation, placing compaction, surface finishing, curing, joints, others* Reinforcement formworks and shoring *cutting and bending formwork, shoring, placing* Factors for PC *grouting, qualification, etc.* Others
subtotal A	subtotal B	subtotal C

Examination of Limit State   total score $= A + B + C > DI_{req}$

Fig.1.   JSCE Durability Design Scheme and Fundamentals.

The authors idea of a versatile method of evaluating durability is shown in Fig.2 and is similar to the dynamic nonlinear FE analysis technique of reinforced concrete, which is the most enhanced way of examining structural and mechanical performance[12]. The emphasis in the proposed method is on identifying and modeling the durability characteristics of concrete at the micro level. If these characteristics are well understood at the micro scale they can be integrated over the domain of interest to predict the durability of concrete over its life span. It is believed that the physio-chemical characteristics of concrete at the micro level are more fundamental in nature and a durability design system based on these characteristics can be seamlessly integrated into the well developed system of examining structural and mechanical performance. The output of the general durability design method is 1) micro pore structure of concrete as expressed by pore size distribution, 2) degree of hydration of cement in concrete, 3) pore water content, 4) internal stress related to shrinkage and external actions and 4) chemistry of pore solution, which is related to the corrosion of steel and the deterioration of the cement paste matrix. The above information is to be computed in 3D space and time domains. The input information is 1) mix proportions of concrete, 2) mineral composition of cement (or kind of cement), 3) geometrical shape and dimensions of structure, 4) initial temperature of concrete at placement and 5) environmental boundary conditions to which the structure is exposed.

It must be said that the above generalized computational method has not yet been established even though previous research could make some contribution to the entire process. But, the authors consider that it is the right time to start establishing the generalized computational method of examining durability performance of structural concrete. Also, we should try to synthesize past scientific endeavor and start necessary studies in line with the scheme shown in Fig.2. This method may mathematically cover the articles treated in the durability design shown in Fig.1 as material and structural detail, but cannot cover those items categorized as construction factors. As a matter of fact, we know construction has a significant influence on the durability performance of structural concrete.

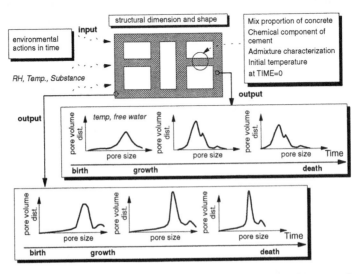

Fig. 2. New proposal for generic verification method of durability performance of concrete.

However, utilizing self-compacting high performance concrete (HPC), which can be placed inside formwork having densely arranged reinforcing bars without any compaction[13], the uncertainties regarding construction methods can be completely avoided in the framework of durability design. Self-compacting HPC was originally developed for realizing reliable structural concrete with high durability when high quality workmanship cannot be expected at the construction site[13]. In other words, HPC eliminates the uncertainties in compaction and construction quality, and makes computational approach meaningful. The authors would like to regard the technology of self-compacting HPC as the basis for the computational approach stated above.

The generalized way of simulating the durability of concrete can be applied to any limit state of durability performance. In this paper, the authors propose an example of the versatility and power of this approach. To meet this challenge, the authors select an engineering problem of early dried concrete structures and its micro pore structure formation, strength development and hydration process.

Owing to the requirements of super fluidity and segregation resistance of self-compacting HPC, a small amount of free water and low water to powder ratio are specified. It leads to very tight texture of micro pores and durability much beyond the required level when sufficient curing is provided. Therefore, under some conditions, the curing procedure can be avoided for making the rapid construction system possible[9]. In this case, it is indispensable to quantitatively evaluate the degradation of durability performance as shown in Fig.3. The following section outlines the scheme for verifying the curing and durability performance in line with the idealized durability examination as shown in Fig.2.

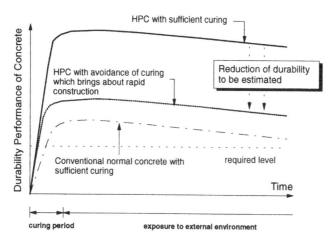

Fig. 3. Evaluation of young concrete subjected to drying, premature hydration and curing.

3 Coupled mass transport, hydration and structure formation theory

In this study the inter-relationships of hydration, moisture transport and pore-structure development processes are analytically defined based upon fundamental physical material models pertaining to each physical process. Physical processes related to moisture transport are formulated at the micro pore scale and integrated over a REV to give macro scale mass transport behavior. The hydration process is based on a multi-component hydration heat model of powder materials and is dependent on the free water available for hydration. Thus the average degree of hydration as well as the hydration of each clinker component can be obtained. The development of the pore structure at early ages is obtained using a pore structure development model based on the average degree of hydration. The predicted computational pore structures of concrete are used as a basis for moisture transport computations. In this way, applying a dynamic coupling of pore-structure development to the moisture transport and hydration models, the development of strength along with moisture content and temperature can be traced with the increase of degree of hydration for any arbitrary initial and boundary condition. This methodology serves as a basis for the quantitative evaluation of parameters relevant for the durability of structural concrete. Interaction between various components is considered at the macro scale. The formulation is kept simple enough to be directly used in a regular FEM code of three dimensions.

3.1 Pore structure development [6]

As a physical basis for pore-structure development computation at early ages of hydration, overall pore space is broadly divided into interlayer, gel and capillary porosity. The powder material is idealized as consisting of uniformly sized spherical particles. After initial contact with water, these particles start to dissolve and the reaction products are precipitated on the outer surfaces of particle and the free pore solution phase. Fig.4 shows a schematic representation of various phases at any arbitrary stage of hydration. Precipitation of the pore solution phase on the outer surfaces of particles leads to the formation of outer products whereas so called inner products are formed inside the original particle boundary where the hydrate characteristics are more or less uniform.

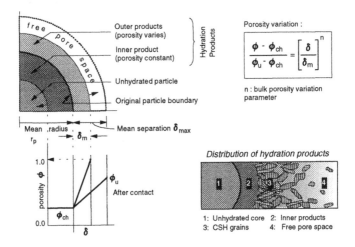

Fig. 4. Hydration of a single cement particle.

Inner product microstructural properties are assumed to be constant throughout the process of pore structure formation. It is assumed that the conventional capillary porosity exists primarily in the outer product whereas the CSH hydrate crystals account for gel and interlayer porosity. Characteristic porosity of the CSH mass ϕ_{ch} is assumed to be constant throughout the progress of hydration. In this study a value of 0.28 is assumed. It has to be noted that this porosity includes both the interlayer as well as micro-gel porosity.

Undertaking these assumptions, weight W_s and volume V_s of gel solids can be computed, provided average degree of hydration α and the amount of chemically combined water β per unit weight of powder material are known. These two parameters are obtained from the multi-component hydration heat model described later. As reported by past researchers, a layer structure for CSH mass is assumed with the interlayer spacing of one water molecule[2,4]. Overall volume balance thus gives interlayer(ϕ_l), gel(ϕ_g) and capillary(ϕ_c) porosity. These parameters are therefore computed as,

$$\phi_l = \frac{t_w s_l \rho_s}{2} \qquad \phi_c = 1 - V_s - (1-\alpha)\frac{W_p}{\rho_p} \tag{3a}$$

$$\phi_g = \phi_{ch} V_s - \phi_l \qquad V_s = \frac{\alpha W_p}{1 - \phi_{ch}}\left(\frac{1}{\rho_p} + \frac{\beta}{\rho_w}\right) \tag{3b}$$

where t_w : Interlayer thickness, s_l : specific surface area of interlayer, W_p : Weight of the powder materials per unit volume, ρ_p : density of the powder material, ρ_s : dry density of solid crystals $= (1+\beta)(1-\phi_{ch})/(\rho^{-1}+\beta\rho_w^{-1})$.

The outer product thickness δ_m is computed by assuming a bulk porosity variation which increases from the characteristic porosity of inner products ϕ_{ch} at the particle surface to unity or ϕ_u at the external boundary of the outer product. It is assumed that before contact occurs between expanding particles, bulk porosity is unity at the outermost boundary of outer products. After contact occurs, bulk porosity at the outermost boundary of the outer product reduces to ϕ_u so as to accommodate new hydration products being formed. Naturally, ϕ_u will keep on decreasing as hydration progresses and new hydration products are formed. With this model of particle expansion during hydration, surface areas of capillary(S_c) and gel pores(S_g) per unit volume of matrix can be obtained as,

$$S_C = \frac{3\delta_m\left[A\delta_m^2 + B\delta_m + C\right]}{l_g(1-\phi_{ch})(r_p + \delta_{max})^3} \qquad S_g = W_s s_g \tag{4}$$

where, l_s : average volume to surface ratio of solid crystals, r_p : Mean powder particle radius, δ_{max} : $R - r_p$, where R denotes equivalent radius of particle cell, s_g : sp. surface area of gel mass, A :$\{n(1-\phi_{ch})+3(1-\phi_u)\}/\{3(n+3)\}$, B : $\{n(1-\phi_{ch})+2(1-\phi_u)\}r_p/(n+2)$, C : $\{n(1-\phi_{ch})+(1-\phi_u)\}r_p^2/(n+1)$. It is implicitly assumed that the inner mass develops inside the original particle boundary with constant properties throughout the hydration process, whereas representative solid crystals of constant and uniform properties are deposited in the outer pore solution phase. With the maturity of the hydrating matrix, microstructural properties tend to become uniform in the outer as well as inner products. Consequently, difference of microstructural properties in the outer and inner products diminishes with hydration. In the computational model, a bi-modal *R-R* porosity distribution is assumed to represent the overall microstructure (combining inner and outer products) which gives total porosity function $\phi(r)$ as

$$\phi(r) = \phi_l + \phi_g\{1 - \exp(-B_g r)\} + \phi_c\{1 - \exp(-B_c r)\} \qquad r : pore\ radius \tag{5}$$

The distribution parameters B_c and B_g can be easily obtained from the computed porosity and surface area values for the capillary and micro-gel region by assuming a cylindrical pore geometry. A generic nonlinear variation of total porosity in the outer products can also be considered in the above model of pore structure development. However, in this study the authors use a linear variation of total porosity in outer products (i.e., n=1).

3. 2 Moisture transport process [5]

The ingress of moisture into the pores of concrete is a thermodynamic process, driven by the pressure and temperature potential gradients. In this study, total water present in matrix pores is subdivided into interlayer, adsorbed and condensed water. Interlayer water is probably under the influence of strong surface forces and does not moves readily under the application of pore pressure potential gradients. Since surface forces can be neglected for pores having radii greater than 5 to 10 *nm*, it appears that most of this fixed water is retained in pores of radii less than 5 *nm*. The amount of water in concrete microstructure in the form of condensed and adsorbed phases can be obtained by integrating the degree of saturation of individual pores over the computed porosity distribution as given by the modified BET equation[3]. Thus, at any given relative humidity the total saturation as a sum of the contributions of water in the gel, interlayer and capillary pores is obtained. It is therefore possible to compute the moisture capacity for any relative humidity, which is defined as the slope of the saturation-RH isotherm at a constant temperature.

Regarding the moisture conductivity of the hydrating matrix, capillary and gel pores are idealized as cylindrical pores. Any liquid movement occurring on the surface of pore walls of the adsorbed phase is also neglected. Liquid transport in the porous network would occur only through the pores which are capillary condensed. Integrating the liquid flux over the complete microstructure using a random pore distribution model, the authors propose, flux q_l of moisture in the liquid phase for a continuous pore distribution as [5],

$$q_l = -\left(\frac{\rho\phi^2\left(\int_{r_m}^{r_k}(r-t_a)\,dV\right)^2}{8\,\mu_a\,\Omega\int_{r_m}^{r_k}dV}\right)\frac{dP_l}{dx} = -K_l\frac{dP_l}{dx} \tag{6}$$

where, ϕ : total porosity (excluding interlayer porosity), r_m : minimum capillary radius = 2.5 nm (IUPAC classification), r_k : radius of the largest filled pore for a given relative humidity, t_a : thickness of adsorbed layer of water[m], $\Omega = (\pi/2)^4$ for a 3-D pore network which is uniformly randomly connected, P_l : pore water pressure which is related to relative humidity by Kelvin's equation [Pa], dV : averaged differential pore volume obtained by porosity averaging of computed capillary and gel porosity distributions, such that $\int_0^\infty dV=1$, μ_a : actual viscosity of water in the porous media = $\mu_i\ exp(\Delta G^*/RT)$ where μ_i : viscosity of water under ideal conditions, ΔG : extra Gibbs free energy for the activation of flow. It has to be remarked here that ΔG seems to be dependent on the moisture history and increases or decreases with the relative humidity in microstructure. A maximum value of 3000 *cal/mol* for ΔG is assumed by the authors based on experimental results. Therefore, the liquid conductivity K_l is dependent not only on the moisture content but on the moisture history as well.

The flux of the the vapor phase is governed by Fick's first law of diffusion. However, factors such as complicated pore network, reduction of pore volume through which vapor movement can take place with increasing saturation, Knudsen diffusion etc. reduce the apparent diffusivity of vapor. To account for the complicated pore network, a random pore model is adopted as in the previous case. Thus, the flux of moisture q_v in the vapor phase can be expressed as

$$q_v = -\frac{\rho_v\phi^2 D_a}{\Omega}\left(\int_{rk}^\infty (1+0.5f/(r-t_a))^{-1}dV\right)\frac{dh}{dx} = -D_v\frac{dh}{dx} \tag{7}$$

where ρ_v : water vapor density [kg/m^3], D_a : vapor diffusivity in free atmosphere [m^2/s], f : mean free path length of water vapor molecule [m], D_v : vapor mass diffusivity in porous medium [kg/m.s], h : relative humidity. Integral in above expression sums up the effect of Knudsen diffusion over the whole pore distribution.

During the early age hydration, substantial amount of free moisture is consumed in the hydration process and gets fixed as chemically combined water of hydration products being formed. Also the mass of the vapor can be neglected compared to the mass of liquid moisture present in the concrete. With these assumptions, the overall moisture balance of concrete can be obtained as

$$\rho\left(\sum\phi_i\frac{\partial S_i}{\partial P}\right)\frac{\partial P}{\partial t} - div\big(K(P,T)\nabla P\big) + \rho\sum S_i\frac{\partial\phi_i}{\partial t} - W_p\frac{\partial\beta}{\partial t} = 0 \tag{8}$$

where, ϕ_i : porosity of each component(interlayer, gel and capillary), S_i : degree of saturation of each component, P : equivalent liquid pore pressure, ρ : density of pore water. Moisture conductivity K is obtained from the flux models explained above, using the computed R-R distribution function. The overall moisture capacity is obtained from the summation of moisture capacity for each component of pore water. The rate of moisture consumption term, $W_p.d\beta/dt$ is dependent on the rate of hydration α, which is in turn dependent on the available free water (taken as capillary condensed water in this model). This inter-dependency of moisture transport and hydration makes the early age hydration problem dynamically coupled.

The general framework can also treat concrete as a composite porous material whereby various components of concrete, namely the hardened cementitious matrix, aggregates and interfaces between aggregate and the matrix can be unified into the moisture transport process. The role and significance of these components in the moisture transport process has been discussed in the past by the authors[5]. Combining the moisture transport processes in these components using the concept of *local moisture transfer*, overall moisture transport behavior can be obtained. However, in this study of early age developments, the local moisture transfer between aggregates and surrounding matrix has not been considered.

3.3 Multi-component cement hydration heat model [7,14]
The hydration process at early ages of concrete is simulated using multi-component model for hydration heat of concrete based on cement mineral compounds. This model assumes a multi-component mineral composition of cement. The hydration process of each clinker mineral present in the cement is combined to represent the overall hydration phenomenon. In view of the concept of a multi-component powder material, the effect of arbitrary types of cement or powder materials can be rationally taken into account to predict the overall heat generation rate. The influence of variable moisture content or free water in the hydrating concrete is also taken into consideration. The total heat generation rate of concrete H per unit volume is idealized as,

$$H = CH_C \qquad H_C = \sum p_i H_i \tag{9}$$

where C is the cement content per unit volume of concrete, H_C is the specific heat generation rate of cement. The clinker components of cement adopted in this model are aluminate(C_3A), alite(C_3S), belite(C_2S), ferrite(C_4AF) and monosulphate. Other powder materials like slag or fly ash can be incorporated in the model as pseudo clinker components. H_i is the specific heat generation rate of individual clinker component, p_i the corresponding mass ratio in the cement such that $\Sigma\ p_i = 1$. The temperature dependent heat generation rate H_i of each clinker component is obtained using Arrhenius's law as

$$H_i = H_{i,T_o} \exp\left[-\frac{E_i}{R}\left(\frac{1}{T}-\frac{1}{T_o}\right)\right] \tag{10}$$

where E_i is the activation energy of i-th component hydration, R is gas constant and H_{i,T_o} is the referential heat rate of i-th component when temperature is T_o. The referential heat rate of each reaction embodies the probability of molecular collision with which hydration proceeds. In the model it is taken as a function of the amounts of free water, the non-dimensional thickness of the cluster made by already hydrated products and unhydrated chemical compounds. The referential heat generation rate is dependent on the amounts of *free* water ω_{free}, the thickness of the cluster η_i made by already hydrated product and unhydrated chemical compound, and on the total accumulated heat of each clinker component. Expressed in a functional form

$$H_{i,T_o} = \beta_i.F(\overline{Q}_i) \qquad \overline{Q}_i \equiv \int \overline{H}_i \tag{11}$$

where, the function F represents the idealized heat generation rate function in terms of total accumulated heat Q_i. The parameter β_i indicates the reduction of hydration rate with respect to the increasing thickness of cluster made of already hydrated product and the decreasing free water during hydration as

$$\beta_i = 1 - \exp\left\{-r.\left(\frac{\omega_{free}}{100.\eta_i}\right)^s\right\} \tag{12}$$

where, constants r,s are material parameter. The amount of free water ω_{free} in above model is in fact the total condensed water in developing micro-structure. Since, the water existing in hydrate crystals cannot be taken up for hydration, this free water for hydration is taken as the water existing in capillary pores in condensed state. The free water for hydration is the primary source of non-linear coupling between moisture transport and hydration heat models. The total amount of water consumed per unit volume of concrete, $W_p.\beta$ due to the chemical reactions with clinker components is incrementally computed at any point of hydration from the usual set of stoichiometric equations of hydration[7]. The hydration heat generation model described above not only provides us with the amount of heat generated in hydration but also the average hydration degree of each clinker mineral as well as the amount of water fixed as chemically bound water in hydration.

Finally, the temperature distribution and the degree of hydration of concrete can be obtained by applying the thermodynamic energy conservation to the space and time domain of interest and using hydration heat generation model as

$$(\rho c)\frac{\partial T}{\partial t} = div(k\nabla T) + H \tag{13}$$

where, k : mean thermal conductivity of concrete, ρc : heat capacity of concrete and H is the concrete heat generation rate computed from various considerations described above (Eqns 9-12).

4 Experimental verification and evaluations

Early age development involves a simultaneous occurrence of hydration, structure development and moisture transport processes described in the previous section. Under moisture sealed curing conditions, most of the hydration process occurs within 2 days after casting. Also a rapid increase of strength gain accompanied by particle to particle contact occurs after few hours of casting.

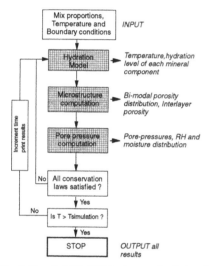

Fig 5. Coupled 3D-FEM scheme of solution for hydration moisture transport and structure formation problem in concrete.

To simulate and verify such processes, a FEM code was written using an explicit coupled scheme of solution to obtain solutions of temperature, pore pressure and other material properties in 3-D space and time domain. The overall scheme of computation is as shown in Fig.5. It has to be noticed that the only input parameters are the W/P ratio, type and composition of powder materials, mix-proportions and environmental conditions specified as initial and boundary conditions which are applied to the model of interest.

Three different case studies were conducted to quantitatively evaluate the coupled computational system of moisture transport, hydration and moisture transport. The first case involves the quantitative study of effect of different curing conditions applied to various mortars, on their strength development. For such a case, the incremental average compressive strength df_c' [MPa] as a function of average degree of hydration α_i of major clinker mineral components of the binder materials was computed [8] as,

$$df_c' = 25dQ_{C_3S} + 40dQ_{C_2S} + 27dQ_{SG} + 40dQ_{FL} \qquad dQ_i = w_i d\alpha_i \qquad (14)$$

where, w_i : ratio of weight of i-th clinker mineral in the powder to mix water, $d\alpha_i$: incremental increase in the degree of hydration of i-th clinker mineral component. The experimental data of strength development and moisture loss with time was obtained for different curing conditions and various mix proportions as shown in Table 1.

Fig.6 shows a comparison of predicted and experimental compressive strength values at 7 and 28 days for mortars of different mix-proportions exposed to various curing conditions as shown in Table 1.

Table 1 : Mix proportions and curing conditions for curing-strength gain experiment [9].

Case	W/C	Air	Mix Unit Weight kg/m³							Case	Curing condition
	(%)	(%)	W	C	MC	Lime	Slag	Sand	G	SL	Sealed
MS	33.5	3.5	172	--	513	28	--	828	827	16	16 hrs stripped
S6	55.8	3.5	172	--	308	17	200	828	827	2D	2 days stripped
OP	55.0	4.5	165	300	--	--	--	927	924	WT	submerged

MC : Medium heat cement; C : Ordinary Portland Cement; G : Gravel

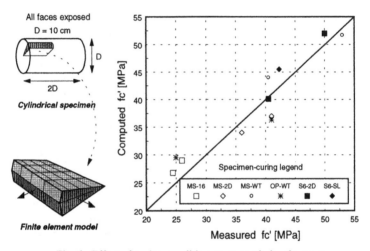

Fig. 6. Effect of curing condition on strength development.

The second case is the prediction of weight loss with time for specimens of different mix proportions and cured for different periods. After a stipulated period of curing period as shown in Fig. 7 the specimens were put in a vacuum desiccator and mass measurements were taken with time[10,11]. Numerical computation using finite-element scheme described in Fig. 5 was also done simulating the experimental procedure to obtain quantitative estimates of various physical quantities related to the early age development and drying behavior of mortars. The computed results show reasonable agreement not only in terms of the rate of moisture loss, but also in terms of the absolute amount of moisture loss of various mortar specimens(Fig.7). However, it must be stated that agreement for cases where the mix was prepared with slag and pozzolans as powder material, does not show very good agreement. Probably the treatment of powder material as single sized particle tends to overestimate the capillary porosity distribution leading to greater conductivity and mass loss.

The third case study is the comparison of the effect of early drying on micro-structure formation and mass loss behavior. Specimens of 0.265 *W/C* ratio were exposed to one-dimensional drying in 40%RH and 20°C after one day moisture sealed

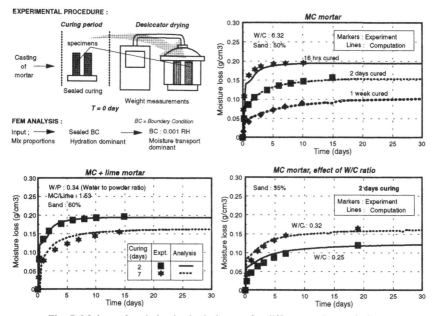

Fig. 7. Moisture loss behavior in desiccator for different curing periods.

curing. After 8 days of drying, microstructure were measured, at 1 cm interval from the exposed surface. Experimental data showed that hydration almost immediately stops near the surface after exposure while the inner parts continue to hydrate. The microstructure formed in the surface zone is markedly different and coarser compared to the inner zones. A numerical simulation using the same initial and boundary condition as the experiment, shows a similar trend and exhibits an extremely strong coupling between hydration and the moisture transport process at early ages. It is observed that while no development of microstructure occurs near the surface due to a rapid loss of moisture, the inner zone of the sample continues to hydrate and significant microstructural development takes place. The computed moisture profiles at two and eight days after drying also show reasonable agreement with the measured moisture profiles (Fig. 8).

5 Conclusions

Using simple physical models for the hydration, moisture transport and microstructure development, early age development process and various parameters relevant for long term durability can be quantitatively obtained. A finite-element based computational model has been developed by combining the pore-structure development, hydration heat and moisture transport models. With this computational model the interdependency of hydration, structure development and moisture transport mechanisms can be simulated in a rational way. The highlights of the various models incorporated into the proposed durability evaluation system are :

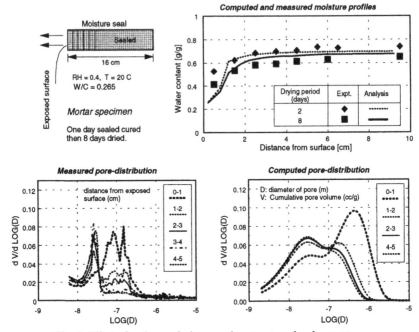

Fig. 8. Effect of early age drying on microstructure development.

1. The pore structure development is dependent on the state of maturity of the hydrating matrix. Also, the pores in the developing microstructure are primarily sub-divided into interlayer, gel and capillary pores.
2. The moisture transport and retention characteristics of hardening concrete have a direct bearing with the developing microstructure and the same are obtained directly from the computed microstructures.
3. Regarding hydration, the cement clinkers are classified into four minerals with which five patterns of hydration are embodied. The hydration rate is directly coupled to the free water and temperature, which represent the thermodynamic environment of cement in concrete.

Overall the combined system has very few empirical parameters and as such can be applied with good confidence to study the cases of different mix proportions under various curing conditions on early age development and related problems. Preliminary verifications have shown a reasonable agreement with experiments for various aspects of early age development. As for future development, required is a combination of the coupled mass transport, hydration and pore structure formation theory with the structural mechanics model of reinforced concrete for examining total performance of reinforced concrete from the birth to death.

The authors express their sincere gratitude to Prof. H. Okamura, the University of Tokyo, for fruitful discussion and suggestions.

6 References

1. JSCE Subcommittee on durability design for concrete structures (1989), Proposed recommendation on durability design for concrete structures, *Concrete Library JSCE*, No. 65.
2. Feldman, R.F. and Sereda, P.J. (1968) A model for hydrated Portland cement paste as deduced from sorption-length change and mechanical properties. *Mater. Constr.*, Vol. 1, 509-519.
3. Hillerborg, A. (1985) A modified absorption theory. *Cem. & Concr. Res.* Vol. 15, 809-816.
4. Powers, T.C. (1965) Mechanism of shrinkage and reversible creep of hardened concrete. *Proc. Int. Symp. Struct. Concr. London*, (Cem. & Concr. Assoc., London), 319-344.
5. Chaube, R.P., Shimomura, T., Maekawa, K. (1993) Multiphase water movement in concrete as a multi-component system, *Proceedings of the Fifth Intn'l ConCreep RILEM Symp*, 139-144.
6. Chaube, R.P. and Maekawa, K. (1995) Coupled moisture transport, structure formation and hydration in cementitious materials, *Proceedings of the JCI*, Vol. 17, No. 1, 639-644
7. Kishi, T. and Maekawa, K. (1993) Multi-component model for hydration heat of concrete based on cement mineral compounds, *Proceedings of the JCI*, Vol. 15, No. 1, 1211-1216.
8. Kato, Y. and Kishi, T. (1994) Strength development model for concrete in early ages based on hydration of constituent minerals, *Proceedings of the JCI*, Vol. 16, No. 1, 503-508.
9. Shimomura, T. and Uno, Y. (1995) Study on properties of hardened high performance concrete stripped at early age, *Proceedings of JSCE*, Vol. 26, No. 508, 15-22.
10. Shimomura, T. (1993) *Drying shrinkage model of concrete based on porosity distribution density function*. Ph.D. Thesis, (In Japanese), The University of Tokyo.
11. Shimomura, T., Fukutome, K. and Maekawa, K. (1995) Analysis of drying shrinkage behavior of concrete by the micromechanical model, *Proceedings of JSCE*, Vol. 27, No. 514, 41-53.
12. Okamura, H. and Maekawa, K. (1991) Nonlinear analysis and constitutive models of reinforced concrete, Giho-do press, Tokyo.
13. Okamura, H., Maekawa, K. and Ozawa, K. (1993) High performance concrete, Giho-do press, Tokyo.
14. Kishi, T., Shimomura, T. and Maekawa, K. (1993) Thermal crack control design of high performance concrete, *Proceedings of International conference on CONCRETE 2000*, Vol.1, E & FN Spon.

9 PERFORMANCE CRITERIA FOR CONCRETE DURABILITY

O.E. GJØRV
The Norwegian Institute of Technology – NTH, Trondheim, Norway

Abstract
Over recent years, an increasing amount of poor long-term performance of concrete structures both represents a big economic problem and a waste of natural resources. In order to approach this problem, integration of conventional structural design and performance based durability design is important. The introduction of relevant performance criteria which can be used both for job specification and control of in situ quality is also important. In the present paper, some quality parameters which can be used as a basis for performance criteria for concrete durability are discussed.
Keywords: Concrete structures, durability, sevice life, performance criteria, quality parameters, test methods, job specifications, quality control.

1 Introduction

In most countries, an increasing amount of poor long-term performance of concrete structures requires large amounts of money and natural resources on repairs and maintenance. There may be several reasons to this situation, but in order to improve the long-term performance, more emphasize should be given to integration of conventional structural design and performance based durability design.

In most concrete codes and job specifications, requirements to concrete durability are almost exclusively based on concrete composition, properties and composition of concrete constituents, casting and compaction procedures as well as curing conditions and compressive strength. For several reasons, this approach has frequently shown to yield insufficient and unsatisfactory results.

In many other industrial areas, performance criteria as well as instructions for operation and preventive maintenance are an integral part of the design procedure.

In order to improve the long-term performance of concrete structures, proper and relevant performance criteria are needed. This is important both from a job specifi-

Integrated Design and Environmental Issues in Concrete Technology. Edited by K. Sakai. Published in 1996 by E & FN Spon, 2–6 Boundary Row, London, SE1 8HN, UK. ISBN 0 419 22180 8.

cation and quality control point of view. In the following, some quality parameters which can be used as a basis for performance criteria for concrete durability are discussed.

2 Water permeation

2.1 Test methods

2.1.1 Water permeability
In the literature, a number of different test methods for determination of water permeability (hydraulic conductivity) is given. In Norway, a test method is frequently used [1], where concrete cylinders or drilled out cores with a diameter of up to 100 mm are embedded in a dense epoxy mortar, from which 50 mm thick slices are cut. After water saturation, the specimens are subjected to a hydraulic gradient equivalent to a 4 MPa water pressure. After a steady-state of water penetration has been reached, the flow of water is observed over a certain period of time, and the permeability coefficient calculated according to Darcy's equation.

2.1.2 Water penetration
In the Norwegian Concrete Code [2], the specified test method for determination of water penetration is based on ISO/DIS 7031. In this test, the concrete specimen is subjected to a three days water pressure, successively increasing from 0.3 MPa to 0.5 MPa and 0.7 MPa over a period of three days. Then, the concrete specimen is split and the depth of water penetration observed.

2.2 Evaluation of test methods

2.2.1 Water permeability
It is generally accepted that information about the permeability is of basic importance to the overall concrete durability. However, it should be noted that the results from such a test are strongly dependent on the particular test conditions used.

Investigations have shown that variation in hydraulic gradient, type of water, temperature and possible evaporation from the output side may substantially affect the

Table 1. Effect of air content in the water at 4.0 MPa hydraulic pressure on the water permeability [1].

Water-cement Ratio	Air content in water %	Water permeability (K) 10^{-12} m/s (%)
0.5	100	1.46 (100)
0.5	33	9.53 (650)
0.6	100	15.7 (100)
0.6	33	29.5 (188)

stady state of water flow [1]. Thus, the data in Table 1 show that when the air content in the water was reduced to approximately 33% of the saturated air content at one atmospheric pressure, the permeability coefficient increased from 1.46 x 10^{-12} to 9.53 x 10^{-12} m/s and from 15.7 x 10^{-12} to 29.5 x 10^{-12} m/sec for the concrete specimens with a water-cement ratio of 0.50 and 0.60, respectively. As the pressure decreases through the concrete specimen, a certain expulsion of air takes place which causes a blocking effect due to released air. This effect depends on the hydraulic gradient (Table 2, Fig. 1). Therefore, when specifications based on water permeability are given, it is important to relate this to a specific test method with properly defined test conditions. For high-quality concrete, testing of water permeability may be a problem. For such concrete, it is frequently observed that the concrete is so dense that it is difficult to get any water through the specimens [3].

Table 2. Effect of hydraulic pressure on the water permeability [1].

Air con-tent in water, %	Water satu-ration	Water permeability (K), 10^{-12} m/s (%)			
		4.0 MPa	3.0 MPa	2.0 MPa	1.0 MPa
100	natural	0.42 (100)	0.36 (82)	0.28 (67)	0.23 (54)
33	natural	2.73 (100)	2.23 (82)	1.86 (68)	2.00 (74)
33	vacuum	2.53 (100)	2.41 (95)	2.14 (85)	2.49 (99)
33	natural	2.97 (100)	2.39 (89)	2.18 (73)	2.03 (68)

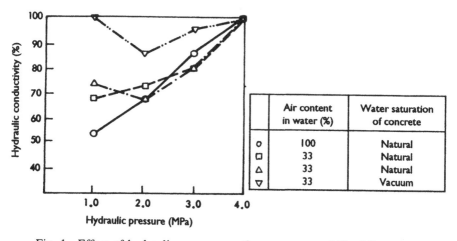

Fig. 1. Effect of hydraulic pressure on the water permeability [1].

2.2.2 Water penetration

While the water permeability is primarily controlled by a hydraulic gradient as the driving mechanism for the water transport, an observed depth of water penetration is the result of the combined effects of both a hydraulic gradient and a water suction.

Since water suction very much depends on the initial moisture conditions within the concrete, proper attention to the test conditions become even more important. Also, when specifications are based on a certain depth of water penetration, it is important to specify whether the requirement applies to a maximum or to an average depth of penetration.

3 Water suction and porosity

3.1 Test methods

3.1.1 Water suction

Water suction or capillary absorption parameters are frequently used as a basis for general assessment of concrete quality. In principle, a relationship between absorbed amount of water and square root of time exists, as shown in Fig. 2. In Norway, a procedure is often used where 20 to 30 mm thick concrete slices are first dried at 105°C to constant weight and then preconditioned according to a certain procedure before exposure to water absorption [4]. Depending on the concrete quality, the water front may need from 2 to 20 hrs to reach the top surface of the specimen.

The slope of the curve in the first stage of water absorption up to the "nick point" in Fig. 2, is defined as the Capillary Number (k):

$$k = \frac{Q_{cap}}{\sqrt{t_{cap}}} \qquad (1)$$

where Q_{cap} is the amount of water absorbed during a period of time t_{cap}, after which a distinct change in rate of water absorption normally is observed. A Capillary Resistance Number (m) is further defined as:

$$m = \frac{t_{cap}}{1^2} \qquad (2)$$

where l is the thickness of the concrete specimen.

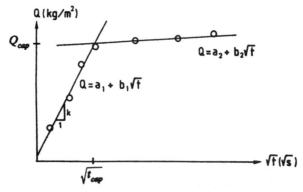

Fig. 2. Relationship between capillary absorption and time.

3.2 Porosity

Based on the same test specimens as that for water suction, a few additional measurements can give further information about the porosity of the concrete.

After the last measurement for capillary absorption, the specimens arecompletely submerged in water for three days, after which the weight A is observed. Then, the specimens are water saturated under high pressure (5 MPa) and the new weight B obtained. The volume V of the specimens are obtained from the different weights observed in water and in air after pressure saturation, respectively. Finally, the dry weight C is determined by drying the concrete at 105°C to constant weight. The following porosity parameters can then be calculated:

Total porosity: $\qquad \varepsilon_{tot} = \dfrac{B-C}{V}$ (%) (3)

Suction porosity: $\qquad \varepsilon_{suc} = \dfrac{A-C}{V}$ (%) (4)

Air content: $\qquad \varepsilon_{air} = \dfrac{B-A}{V}$ (%) (5)

Frost protection number: $\qquad PF = \dfrac{\varepsilon\,air}{\varepsilon\,tot}$ (%) (6)

For this type of testing, any geometrical shape of the concrete specimens can be used. The largest thickness of the concrete specimens should, however, not exceed 30 mm, somewhat depending on the maximum aggregate size and density of the concrete.

3.3 Evaluation of test methods

3.3.1 Water suction

For a given concrete, the Capillary Resistance Number and the Capillary Number are interrelated quality parameters. Both of them are related to the time t_{cap} when a change in rate of water absorption takes place. Both of them reflect the size distribution and continuity of the capillaries. However, since the Capillary Number also depends on the cement paste content which is often not known, the Capillary Resistance Number is generally considered to be a more appropriate quality parameter.

For the testing of water suction parameters, attention to the initial moisture conditions within the concrete is important, since this very much will influence the measurements. If the concrete is only partially dried before testing, experience indicates that it may be difficult to control the distribution of the remaining moisture content within the concrete specimen. Also, if the initial moisture conditions vary from one type of concrete to another, the suction parameters may not be comparable. If the initial moisture content is too high, investigations have also shown that the capillary absorption may not necessarily follow the "square-root-of-time-law" as shown in Fig. 2 [4]. This is the reason why a complete drying at 105°C to constant weight is often used. Although such a severe drying may affect the microstructure of the concrete [5], it is then assumed that the coarser capillary pores which primarily control the water absorption up to the "nick point", is not much affected. However, microcracking will normally occur, the extent and effect of which will then be part of the general quality assessment.

3.3.2 Porosity

Characteristic differences in the pore system can often explain a difference in durability from one type of concrete to another. In addition to the total pore volume (ε_{tot}), it is important, therefore, to know how much of the pore volume which can be filled up with water from an outside source (ε_{suc}). The remaining part of the pore volume, which does not become water filled (ε_{air}), mainly reflects information about the continuity of the pore system. The ratio between (ε_{air}) and (ε_{tot}) or the so-called PF number also reflects the resistance of the concrete to withstand freezing and thawing [6].

4   Chloride penetration

4.1   Test methods

4.1.1 Chloride permeability

Since the "Rapid Chloride Permeability Test" (AASHTO T277-83) [7] was introduced in the USA in the early 1980's, this method has increasingly been applied in several countries. However, since this method is only based on the measurement of the total charge passed through the concrete specimen over a short period of time, it does not give any specific information about the resistance of the concrete against chloride penetration. In Norway, therefore, a similar method was introduced [8], where the measurement was based on the observed rate of chloride penetration through the concrete. Since the beginning of the 1980's, this method has been widely used in Norway for general quality evaluation of concrete durability [9-13].

Based on the same experimental setup as that for the "Chloride Permeability Test" (Fig. 3) and the observed steady-state rate of chloride penetration, the chloride permeability (K_{cl}) can be calculated on the basis of the following equation:

$$K_{cl} = \frac{VL}{Ac_o} \frac{dc}{dt} \quad (cm^2/s) \tag{7}$$

where   V   = volume of the chloride collecting cell (cm^3)
L   = length of concrete specimen (cm)
A   = cross sectional area of specimen (cm^2)
c_o = initial chloride concentration in the chloride source
solution ($mmol/cm^3$)
dc/dt = rate of chloride penetration ($mmol/cm^3 s$)

This test method which is still widely used, is based on a 3 % (approximately 0.5 M) NaCl solution in the chloride source cell, where also the negatively charged electrode is placed, while the chloride collecting cell contains a solution of 0.3 M

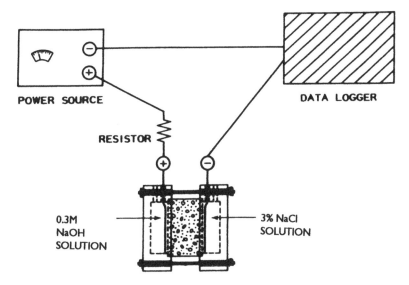

Fig. 3. Experimental setup for testing of chloride permeability.

NaOH, in which the positively charged electrode is placed. A power supply of 12±1 V is applied, and the test is run until a steady-state rate of chloride penetration is obtained. All concrete specimens are subjected to vacuum saturation before testing. Also, both before and after testing, the electrical resistivity is observed as additional information about the concrete quality.

4.1.2 Chloride diffusivity

The conventional testing of chloride diffusivity of concrete is based on immersion of concrete specimens in various types of salt solution. Often, concrete cylinders with a diameter of up to 100 mm are first water saturated and then painted with a dense surface coating or embedded in a dense epoxy mortar in such a way that only one end of the specimen is exposed to the salt solution. This salt solution may be a given concentration of one type of salt or a combination of different types of salt such as seawater. After a certain period of exposure, the profile of chloride penetration is determined on powder samples successively ground off at various depths from the exposed surface. Based on the chloride profile obtained, the effective diffusion coefficient is calculated according to Fick's second law [14] for semi-finite diffusion:

$$c_x = c_o [1 - erf \frac{x}{2\sqrt{D_e\, t}}] \qquad (8)$$

with the following boundary conditions:

$$c_x = 0 \text{ at } t = 0, 0 < x < \infty$$
$$c_x = c_o \text{ at } x = 0, 0 < t < \infty$$

where c_x = chloride concentration at distance x
  c_o = surface chloride concentration

x   = distance from the exposed surface
t   = time
D_e = effective diffusion coefficient
erf   = Gaussian error function

In order to obtain an appropriate basis for the above calculation, the curve fitting for the chloride profile should be based on at least eight individual observations. The values c_s and D_e are then obtained from a non-linear curve fitting.

One of the shortcomings of this type of testing is the long period of testing time normally involved. In order to use chloride diffusivity as a more general parameter for evaluation of concrete quality, it has been an increasing interest to apply various types of accelerated test methods. One approach is to increase the concentration of the chloride source solution [15], but in this case, the diffusion behavior may no longer follow Fick's law for diffusion due to the strong effect of ionic interaction [16]. Another approach is to apply an external electrical field for accelerating the rate of chloride penetration.

In Sweden, an accelerated, non-steady state method was recently introduced [17], where the actual chloride penetration is observed, either in the form of total depth or profile of penetration. If the total depth of chloride penetration is observed, the chloride diffusivity can be calculated according to the following equation:

$$D = \frac{zFE}{RT} \frac{(x_d - \alpha x_d^\beta)}{t} \quad (cm^2/s) \qquad (9)$$

where   D   = diffusion coefficient (cm²/s)
          z   = ion valence in the testing chloride salt
          F   = Faraday constant (J/V·mol)
          E   = electrical field (V/m)
          R   = gas constant (J/K·mol)
          T   = temperature (K)
          x_d = chloride penetration depth (mm)
          α, ß   = regression constants
          t   = duration of testing (hr)

In Norway, an accelerated, steady-state method for determination of diffusivity was also recently introduced [18]. Here, the rate of chloride penetration is determined as already described for chloride permeability, but the chloride diffusivity is calculated according to the following equation:

$$D = \beta_o \frac{300 \ kT}{ze_o \Delta\psi} \frac{LV}{c_o A_o} \frac{dc}{dt} \quad (cm^2/s) \qquad (10)$$

where   D   = diffusion coefficient (cm²/s)
          β_o = correction factor for ionic interaction
          k   = Boltzman constant (1.38×10^{-16} ergs/K/ion)
          T   = temperature (K)

z	= ion valence in the testing chloride salt
e_o	= charge of proton (4.8 x 10^{-10}e.s.u.)
$\Delta\Psi$	= applied electrical potential (V)
L	= specimen thickness (cm)
V	= volume of chloride collecting cell (cm^3)
c_o	= initital chloride concentration in chloride source solution ($mmol/cm^3$)
A	= cross sectional area of specimen (cm^2)
dc/dt	= steady -state migration rate of chloride ions ($mmol/cm^3 \cdot s$)

It should be noted that the stronger the chloride concentration is in the chloride source solution, the more retarded is the chloride penetration. This effect which is due to ionic interaction [19], is the reason why a correction factor (β_o) depending on the concentration of the chloride source solution, has been introduced. Thus, for an increasing chloride concentration from 0.1 to 0.5 M NaCl, β_o increases from 1.22 to 1.70.

4.2 Evaluation of test methods

4.2.1 Chloride permeability
Since the testing of water permeability for dense concretes with very low water-binder ratios represents a problem, testing of chloride permeability or migration testing has shown to be more convenient for quality assessment of even very dense, high-quality types of concrete. Based on extensive testing, an empirical relationship between chloride permeability and concrete quality has been obtained as shown in Table 3.

Table 3. Relationship between chloride permeability and concrete quality [20].

Chloride permeability K_{cl} (cm^2/s x 10^{-7})	Concrete quality
> 50	Low
10-50	Moderate
5-10	High
< 5	Very high

It should be noted, however, that also the so-called chloride permeability is an empirical quality parameter, which is strongly dependent on the particular test conditions used. Therefore, chloride diffusivity appears to be a more well defined quality parameter for characterization of the resistance against chloride penetration from a given chloride solution. The chloride diffusivity may even be used in models for prediction of chloride penetration and hence also for evaluation of service life.

4.2.2 Chloride diffusivity
Depending on environmental conditions, chloride ions may penetrate concrete through various mechanisms such as diffusion, permeation and capillary suction. In general, capillary suction may dominate the penetration through a surface layer of the concrete

which is very porous and only partly water saturated. However, if the porosity is very low or the concrete has a high moisture content, a diffusion mechanism may dominate the chloride penetration.

Conventional diffusion testing has shown that the presence of other types of ions in the system both outside and inside the concrete, also is important for the rate of chloride penetration [21-23]. Thus, a change in type of cation from sodium to calcium in the chloride source solution may increase the rate of chloride penetration quite substantially [21]. Therefore, diffusion coefficients obtained from different types of salt solution are not comparable. Also, diffusion coefficients obtained from a solution of a single salt such as NaCl can not be used for assessing the chloride penetration into concrete from a more complex salt solution such as seawater [22].

Based on a standard salt solution, however, the diffusivity obtained can be used as a general quality parameter both for job specification and control of in situ quality. If a diffusion coefficient is specified as part of the quality requirements, migration testing appears to provide both a quick and appropriate basis, not only for evaluation of the concrete resistance against chloride penetration, but also for a more general evaluation of the concrete durability.

5 Concluding remarks

In order to ensure adequate durability and long-term performance of reinforced concrete structures exposed to aggressive environments, relevant quality parameters are needed, which can provide a better basis both for job specification and control of in situ quality. When performance criteria based on quality parameters such as those discussed in the present paper are enforced, a more controlled concrete durability can be obtained.

6 References

1. Gjørv, O.E. and Løland, K.E. (1980) *Effect of Air on the Hydraulic Conductivity of Concrete, Durability of Building Materials and Components*, ASTM STP 691, ed. by P.J. Sereda and G.G. Litvan, pp. 410-422.
2. *Norwegian Standard* NS 3420 (1989), Oslo.
3. Zhang, M.-H. and Gjørv, O.E. (1991) *Permeability of High-Strength Lightweight Concrete*, ACI Materials Journal, Vol. 88, No. 5, pp. 463-469.
4. Smeplass, S. (1988) *Capillary absorption as a quality criteria*, Report STF6ST A88028, SINTEF, Trondheim, 50 p (in Norwegian).
5. Litvan, G.G. (1976) *Variability of the Nitrogen Surface Area of Hydrated Cement Paste*, Cement and Concrete Research, Vol. 6, pp. 139-143.
6. *Finnish Standard* SFS 4475 (1988) Concrete Frost Resistance, Helsinki.
7. Whiting, D. (1981) *Rapid Determination of the Chloride Permeability of Concrete*, Report No. FHWA/RD-81/119, Portland Cement Association.
8. *Nordtest Method* (1989) NT BUILD 355, ISSN 0283-7153.
9. Detwiler, R.J., Kjellsen, K.O. and Gjørv, O.E. (1991) *Resistance to Chloride Intrusion of Concrete Cured at Different Temperatures*, ACI Materials Journal, Vol. 88, No. 1.

10. Sandvik, M. and Gjørv, O.E. (1992) *High Curing Temperatures in Lightweight High-Strength Concrete,* Concrete International, Vol. 14, No. 12, pp. 40-42.

11. Gjørv, O.E. and Martinsen, J. (1993) *Effect of Elevated Curing Temperature on High-Strength Lightweight Concrete*, 3. International Symposium on Utilization of High-Strength Concrete, Lillehammer, Norway, Proceedings, pp. 706-712.

12. Gjørv, O.E., Tan, K. and Zhang, M.-H. (1994) *Diffusivity of Chlorides from Seawater into High-Strength Lightweight Concrete*, ACI Materials Journal Vol. 91, No. 5, pp. 411-416.

13. Gjørv, O.E., Tan, K. and Monteiro, P.J.M. (1994) *Effect of Elevated Curing Temperature on the Permeability of High-Strength Lightweight Concrete*, Cement, Concrete and Aggregate, CCAGPD, Vol. 16, No. 1, pp. 57-62.

14. Crank, J. (1975) *The Mathematics of Diffusion*, Oxford University Press, 2nd Edition.

15. Sørensen, H. and Fredriksen, J.M. (1990) *Testing and Modelling of Chloride Penetration into Concrete*, Nordic Concrete Research, Research Projects 1990, Oslo, pp. 354-356.

16. Zhang, T. and Gjørv, O.E. *Diffusion Behavior of Chloride Ions in Concrete*, Cement and Concrete Research (paper submitted).

17. Tang, L. and Nilsson, L.-O. (1992) *Rapid Determination of the Chloride Duffusivity in Concrete by Applying an Electrical Field*, ACI Materials Journal, Vol. 89, No. 1, pp. 49-53.

18. Zhang, T.W. and Gjørv, O.E. (1994) *An Electrochemical Method for Accelerated Testing of Chloride Diffusivity in Concrete*, Cement and Concrete Research, Vol. 24, pp. 1534-1548.

19. Zhang, T. and Gjørv, O.E. (1995) *Effect of Ionic Interaction in Migration Testing of Chloride Diffusivity in Concrete*, Cement and Concrete Research Vol. 25, No. 7, pp. 1535-1542.

20. Division of Building Materials, The Norwegian Institute of Technology, NTH, Trondheim (unpublished results).

21. Gjørv, O.E. and Vennesland, Ø. (1987) *Evaluation and Control of Steel Corrosion in Offshore Concrete Structures*, Concrete Durability, Katharine and Bryant Mather International Conference, ACI SP-100, Vol. 2, ed. by J.M. Scanlon, pp. 1575-1602.

22. Gjørv, O.E. and Vennesland, Ø. (1979) *Diffusion of Chloride Ions from Seawater into Concrete*, Cement and Concrete Research, Vol. 9, pp. 229-238.

23. Feldman, R.F. (1986) *Pore Structure, Permeability and Diffusivity as Related to Durability*, 8. International Congress on the Chemistry of Cement, Rio de Janeiro, Vol. 1, pp. 336-356.

24. Ushiyama, H. and Goto, S. (1974) *Diffusion of Various Ions in Hardened Portland Cement Pastes*, 6. International Congress on the Chemistry of Cement, Moscow.

25. Ushiyama, H., Iwakakura, H. and Fukunaga, T. (1976) *Diffusion of Sulphate in Hardened Portland Cement*, Cement Association of Japan, Review of 30th General Meeting, pp. 47-49.

10 MICROSTRUCTURE AND MASS TRANSPORT IN CONCRETE

R.J. DETWILER
Construction Technology Laboratories, Skokie, Illinois, USA

Abstract

With the deterioration of the infrastructure, the durability of concrete is a topic of increasing concern. Since most deterioration mechanisms involve the ingress of water and/or other harmful materials into the concrete, its transport properties are closely related to its durability. Both permeability and diffusivity are greatly increased once clear pathways through the material are established. These pathways can be observed on both the macro and micro scales. This paper reviews the effects of water/cement ratio, curing, and supplementary cementing materials on the microstructure and transport properties of concrete. Permeability and diffusivity can be minimized by a low water/cement ratio, extended curing times, low curing temperatures, and the use of supplementary cementing materials. Where elevated curing temperatures cannot be avoided, supplementary cementing materials can significantly improve the microstructure and transport properties of the concrete. However, since different dosages and combinations of supplementary cementing materials can yield dramatically different results, the mix design must be optimized for the curing regime to be used.
Keywords: Concrete, chloride ions, diffusion, durability, mass transport, microstructure, permeability, porosity, water.

1 Introduction

Concrete is a porous material. As such, its properties are determined not so much by the nature of its solid material as by its pores. Structural engineers are familiar with the effect of a flaw on the measured strength of a specimen. In 1920 Griffith [1] explained why the theoretical strength of a material (calculated from the strength of the bond between individual pairs of atoms) is an order of magnitude greater than the measured strength of specimens of that material: flaws, pores, and cracks concentrate stresses locally and cause the specimen to fail. Subsequent research in fracture mechanics quantified the effects of the geometry of flaws on the failure load in a structural member.

Integrated Design and Environmental Issues in Concrete Technology. Edited by K. Sakai. Published in 1996 by E & FN Spon, 2–6 Boundary Row, London, SE1 8HN, UK. ISBN 0 419 22180 8.

Pores and other voids are likewise disproportionate in their effect on durability. There are many different mechanisms by which concrete can be damaged by its environment. For the most part it is not possible to prevent deterioration: if concrete is kept wet and subjected to repeated cycles of freezing and thawing, for example, it will eventually fail. The purpose of designing for durability is to limit the rate of deterioration, and one of the best ways to do so is to confine the deterioration mechanism to as near the concrete surface as possible. Many deterioration mechanisms involve the ingress of a harmful substance from the environment into the concrete. Thus it is essential to keep the permeability (the ease with which liquids or gases under pressure can move through a porous material) and diffusivity (the ease with which dissolved solids can move through a saturated or partially saturated porous material) of the concrete to those substances to a minimum.

Most damage mechanisms are affected by the permeability of the concrete. Oxygen, whether present in the air or dissolved in water, is one of the reactants in the corrosion of steel. (Steel can corrode without oxygen, but the rate is much slower.) Carbon dioxide combines with water and the calcium hydroxide in the hydrated cement in a reaction called *carbonation*, which lowers the pH and makes the steel more vulnerable to corrosion. Water leaches out calcium hydroxide, making the matrix more porous and more permeable. It also expands on freezing, causing potentially disruptive stresses. Aside from being harmful in itself, water also carries other harmful species:

- Acids that dissolve the calcium hydroxide.
- Sulfate that reacts expansively with the product of the hydration of C_3A (tricalcium aluminate), one of the components of cement.
- Chloride ions, which act as catalysts to the corrosion reaction.
- Alkalies, which react expansively with some aggregates.
- Carbon dioxide.
- Magnesium ions, which substitute for the calcium and leave a weak material.

Water also participates directly in some deterioration reactions; even if the other reactants are all present, the reactions will not proceed unless water is present.

- Alkali-silica reaction.
- Corrosion of reinforcement.
- Sulfate attack.

Thus where durability is a concern, low permeability and/or diffusivity are necessary in order to keep harmful substances (usually water among them) *out*. For certain applications, however, low permeability and diffusivity are also necessary to keep harmful substances *in*, namely hazardous (radioactive and/or toxic) wastes.

2 "Macro" permeability

Although the specific values of permeability and diffusivity for a given concrete vary depending on the substance that is moving through it, the principles for obtaining low values are the same. It is convenient to divide the concept of permeability into two parts: "macro" permeability and "micro" permeability. "Macro" permeability is largely in the hands of structural engineers, who may not even be aware of the effect of their designs and specifications on the durability. Cracking is probably the biggest single factor in the overall permeability and durability of a structure which is exposed to a harsh environment. From a durability standpoint an ideal concrete would have no cracks whatsoever; in practice one should minimize the width and depth of the cracks. Of course ultimately all cracking occurs because the local tensile stress exceeds the local tensile strength of the material.

Causes of stress include plastic shrinkage, drying shrinkage, thermal gradients, and applied loads. In addition, cold joints can act like cracks as far as the permeability and durability are concerned.

Good drainage is essential to the durability of a structure or pavement. Water that splashes onto the surface and then drains away will have little opportunity to penetrate into the concrete. Water that ponds on the surface and remains for hours or days will be able to penetrate farther, carrying with it whatever harmful substances are present. Control joints should be located so that water will not drain into them. Detailing to provide adequate cover will lengthen the time before chloride ions penetrate to the reinforcement. In addition, the configuration of the forms and the spacing of the reinforcement should be such as to allow good consolidation.

Only when proper attention has been paid to the "macro" aspects of permeability does it become worthwhile to examine the "micro" aspects. This is where the pore structure becomes significant. Porosity, which refers to the volume of pores as a fraction of the total volume of a material, is closely related to permeability, but the two are not equivalent. Permeability increases dramatically once a continuous pathway is available, so connected pores contribute much more to permeability than even large discrete pores. Thus it is essential to minimize bleeding, which creates channels under aggregate particles and through the matrix. Mix design factors such as proportions, aggregate grading, and water/cement ratio all affect the bleeding.

3 "Micro" permeability and diffusivity

3.1 Transition zone

Mehta [2] considers concrete to be a three-phase material comprising aggregates, hydrated cement paste, and an interfacial or "transition" zone between the other two phases. Particularly in the early stages of the hydration of the cement, the nature of the transition zone is distinctly different from that of the bulk cement paste. The accumulation of bleed water under the aggregate particles creates local regions of high porosity. Large crystals of ettringite and calcium hydroxide grow in the transition zone. The calcium hydroxide crystals, which have the form of stacks of hexagonal plates, tend to grow parallel to the surface of the aggregate particle. These crystals are easily cleaved between the plates. Differential volume changes due to temperature change or shrinkage due to loss of water in the paste may cause microcracking in the weak, porous transition zone even before loads are applied to the structure. Thus the transition zone is generally considered the "weak link" in concrete.

Even when no bleeding occurs, the transition zone is more porous than the bulk cement paste due to the "wall effect." Bache [3] explains the wall effect in terms of particle packing: near the surface of an aggregate particle, the cement grains cannot pack so closely as they do in the bulk cement paste, as shown in Figure 1. Bache attributes the superior performance of silica fume concretes in part to the improved particle packing that can be attained when the extremely fine silica fume particles help to fill the voids between the cement grains.

Aside from weakening the concrete from a structural point of view, the transition zone also serves as a conduit for the ingress of harmful materials. Halamickova et al. [4] studied the effect of sand content on the water permeability and chloride ion diffusivity of mortars cured to varying degrees of hydration as determined by the content of nonevaporable water. Figure 2 shows their results for the water permeability tests. The increase in permeability with increasing water/cement ratio can be explained in terms of an overall increase in porosity. Continued hydration results in a reduction in total pore volume and a division of

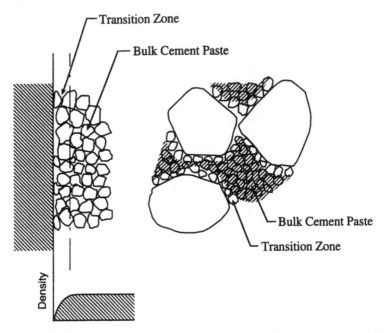

Figure 1. Particle packing of cement grains near the surface of an aggregate particle is less dense than in the bulk cement paste due to the wall effect (adapted from Bache [3]).

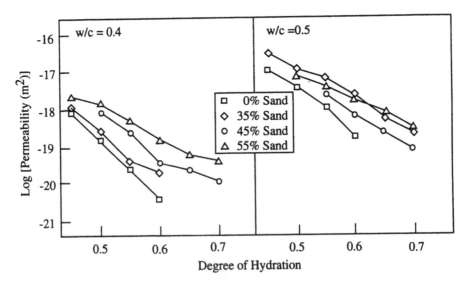

Figure 2. Water permeability as a function of degree of hydration for 0.40 and 0.50 water/cement ratio mortars. The degree of hydration was determined from the nonevaporable water content (Reprinted from Halamickova et al. [4] with kind permission from Elsevier Science Ltd., The Boulevard, Langford Lane, Kidlington OX5 1GB, UK).

large pores into smaller ones having less continuity. An increase in sand content (here expressed as % by volume) introduces more interfaces. In this work the specimens were rotated during the first 24 hours of hydration, so bleeding was not a factor. Similarly, the sand content affected the apparent diffusion coefficients for chloride ions. Table 1 shows some of their data for the 0.50 water/cement ratio specimens.

Table 1. Apparent diffusion coefficients for chloride ions through 0.50 water/cement ratio paste and mortars (Halamickova et al. [4]).

Sand (% by volume)	D (m^2/s x 10^{-11}) 55% hydration
0	2.96
45	5.66
55	5.3

3.2 Curing

In addition to the total degree of hydration, the curing temperature has a significant effect on the pore structure and thus the transport properties of cement and concrete. It has long been known that elevated curing temperatures, while accelerating the early strength gain, result in lower ultimate strengths in concrete. Based on the indirect evidence of heat of hydration and X-ray diffraction experiments, Verbeck and Helmuth [5] explained that at low temperatures, the hydration products have sufficient time to diffuse throughout the cement paste matrix and precipitate uniformly. At higher temperatures, the rate of hydration is so much faster than the rate of diffusion that most of the hydration products remain near the cement grains, leaving the interstitial space relatively open. They believed that the relatively dense depositions of hydration products surrounding the cement grains could serve as diffusion barriers to water and hydration products, thus hindering hydration at later stages.

Using backscattered electron imaging, Kjellsen et al. [6] obtained direct evidence that portland cement pastes hydrated at 50°C had a much more open pore structure than companion specimens hydrated to the same degree of hydration at 5°C. Figure 3 shows backscattered electron images of 0.50 water/cement ratio pastes hydrated at 5, 20, and 50°C to a degree of hydration of approximately 70%. To a first approximation, backscattered electron images can be viewed as plots of the density of the material, with the most dense material appearing the lightest in color. Thus in these images the unhydrated cement grains appear brightest, the calcium hydroxide light gray, other hydration products darker gray, and the pores black. In the specimens hydrated at 20 and 50°C, dense "shells" of hydration products are visible around the cement grains, as predicted by Verbeck and Helmuth. In the specimen hydrated at 50°C, it is relatively easy to trace clear pathways such as would be used by water or ions moving throughout the material.

In a later study, the same authors [7] found that the main difference in pore size distribution as measured by mercury intrusion porosimetry was in the volume of pores of radius 200-1000 Å. They observed that this size of pores coincides with the division proposed by Mehta and Manmohan [8] between "large" (> 500 Å radius) pores which are primarily responsible for the permeability and the smaller pores which have little effect. Detwiler et al. [9] then investigated whether their observations of the effects of curing temperature on the microstructure of cement paste could be observed as effects on the transport properties of concrete. They made concretes with 0.40, 0.50, and 0.58 water/cement ratio and cured them at the same temperatures they had used for the paste specimens. Curing times were allowed to vary to give approximately equal degrees of

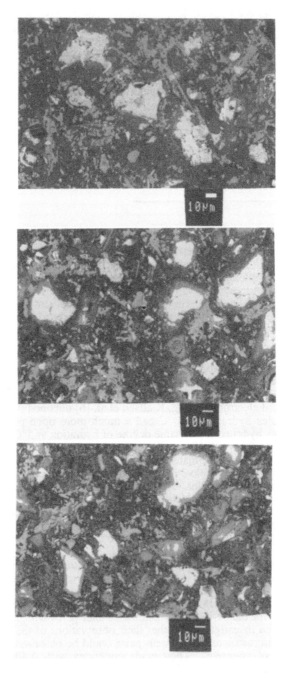

Figure 3. Backscattered electron images of 0.50 water/cement ratio pastes hydrated at 5 (top), 20 (middle), and 50°C (bottom) to a degree of hydration of approximately 70% (Reprinted from Kjellsen et al. [7] with kind permission from Elsevier Science Ltd., The Boulevard, Langford Lane, Kidlington 0X5 1GB, UK).

hydration. They measured the rate of diffusion of chloride ions through these concretes using a modified version of AASHTO T 277 in which the electrical potential is reduced to 12 v and the concentration of chloride ions in the solution at the downstream face of the specimen is measured directly. They found that although there was little difference between the specimens cured at 5 and 20°C, the specimens cured at 50°C had much higher rates of diffusion. They also found that the effect of raising the curing temperature from 20 to 50°C was roughly the same as that of raising the water/cement ratio from 0.40 to 0.50. They concluded that for concrete structures exposed to marine or other corrosion-inducing environments, curing temperature is critical to durability. They suggested the possibility that concretes containing fly ash, silica fume, or slag might perform more satisfactorily under these conditions.

3.3 Supplementary cementing materials

Detwiler et al. [10] used two different methods to assess the resistance to chloride ion intrusion of concretes cured at constant temperatures of 23, 50, and 70°C to approximately equal degrees of hydration. In this work the degree of hydration was determined from the nonevaporable water content as before, except that the "100%" hydration values first had to be established for these materials in these combinations, and then the time required to reach the desired degree of hydration determined for each combination of water/cement ratio, cementitious material, and hydration temperature. Fapohunda [11] gives the details of this procedure. They found that although all concretes were adversely affected by elevated curing temperatures, both silica fume and slag were effective in reducing the rate of chloride ion diffusion in the concretes cured at any given temperature. They also repeated the tests on the same concrete mixes cured according to an 18-hour steam curing regime typical of that used in the fabrication of precast concrete elements (maximum temperature 65°C), finding that both slag and silica fume reduced the charge passed in six hours in the AASHTO T 277 test. They concluded that for any given curing condition, the use of either silica fume or slag has a far more pronounced effect on chloride diffusion than lowering the water/cement ratio from 0.50 to 0.40.

Campbell and Detwiler [12] tested a series of steam-cured concretes having the same water/cementitious materials ratio and cementitious materials content. They selected a water/cementitious materials ratio of 0.45, since that is the maximum allowed by ACI and CSA for concretes exposed to aggressive environments. They subjected the concretes to the same steam curing regime described above, then immediately began conditioning the specimens for the AASHTO T 277 test, so that the concretes were tested at age 48 hours. Their results are shown in Table 2. Both slag and silica fume proved to be effective in reducing the diffusivity of chloride ions, with the combinations of slag and silica fume most effective. Note that all of the concretes tested meet the requirements of ACI and CSA for concretes exposed to aggressive environments, but their performance in service could be expected to be very different.

Cao and Detwiler [13] studied the microstructure of cement paste companion specimens to the concretes tested by Detwiler et al. [10] and hydrated at the same temperatures. Figure 4 shows backscattered electron images of 0.50 water/cementitious materials ratio pastes containing portland cement only, 5% silica fume, and 50% slag, all hydrated at 70°C to a degree of hydration of approximately 70%. Both the silica fume and slag pastes have more discontinuous pore structures than the portland cement paste cured under comparable conditions. The authors also found that all of the specimens cured at 70°C were more porous and less uniform than comparable specimens cured at 23°C to the same degree of hydration. However, the presence of either silica fume or slag mitigated the negative effect of the elevated curing temperature. These results are consistent with the transport properties of the concretes cured at elevated temperatures.

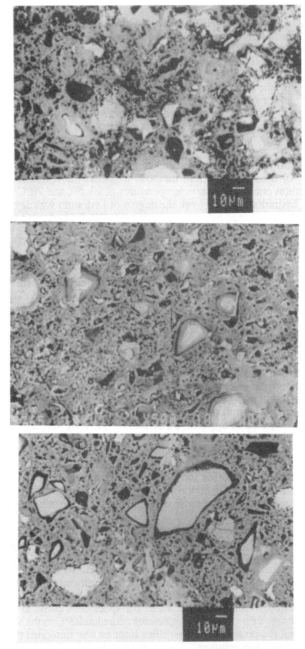

Figure 4. Backscattered electron images of 0.50 water/cementitious material ratio cement pastes hydrated at 70°C to a degree of hydration of approximately 70%. The top image shows a portland cement paste, the middle image a paste containing 5% silica fume, and the bottom image a paste containing 30% slag (Reprinted from Cao and Detwiler [13] with kind permission from Elsevier Science Ltd., The Boulevard, Lanford Lane, Kidlington 0X5 1GB, UK).

Table 2. Results of AASHTO T 277 for steam-cured concretes having water/cementitious materials ratio 0.45 (Campbell and Detwiler [12]).

Concrete	Total charge passed (coulombs)	Rating
Control	11,130	High
30% slag	7800	High
40% slag	7690	High
50% slag	4500	High
5% silica fume	1780	Low
7.5% silica fume	910	Very low
10% silica fume	290	Very low
30% slag + 7.5% silica fume	350	Very low
40% slag + 7.5% silica fume	200	Very low
30% slag + 10% silica fume	150	Very low

4 Conclusions

The permeability and diffusivity of concrete can be related to the microstructure. Specifically, continuous pathways in the form of cracks and pores contribute to the ease with which gases, liquids, and ions move through the material. Since most of the deterioration mechanisms are related to the ingress of water or other harmful materials into the concrete, a favorable microstructure is essential to durable concrete. Transition zones around the aggregate particles serve as pathways for the ingress of water and ions because these regions are more porous than the bulk cement paste. However, extended curing times allow these regions to be filled with hydration products so that they become less porous. A low water/cement ratio minimizes the initial porosity which must be filled with hydration products if the concrete is to have low permeability.

Elevated curing temperatures can be deleterious to the microstructure and transport properties of concrete. At elevated temperatures, the rate of hydration of the cement is much faster than the rate of diffusion of the hydration products, so that the hydration products form dense shells immediately surrounding the cement grains, leaving the interstitial space relatively open. The effect of curing temperature on the diffusivity to chloride ions can be compared to that of significantly increasing the water/cement ratio. Alternatively, the benefits of specifying a high quality concrete can be eliminated by high curing temperatures.

The use of supplementary cementing materials can significantly improve the transport properties and durability of concrete which must be cured at elevated temperatures. Both slag and silica fume contribute to a refinement of the pore structure, leaving fewer clear pathways for the ingress of harmful materials. However, different dosages and combinations of supplementary cementing materials can yield dramatically different performance. Thus the mix design should be optimized for the curing regime to be used.

5 References

1. Griffith, A.A. (1920) The Phenomena of Rupture and Flow in Solids. *Philosophical Transactions of the Royal Society of London*, Series A, Vol. 221. pp. 163-198.
2. Mehta, P.K. (1986) *Concrete: Structure, Properties, and Materials*, Prentice-Hall, Englewood Cliffs, NJ.

3. Bache, H.H. (1981) Densified Cement/Ultra-Fine Particle-Based Materials. Presented at the Second International Conference on Superplasticizers in Concrete, Ottawa, Canada.
4. Halamickova, P., Detwiler, R.J., Bentz, D.P. and Garboczi, E.J. (1995) Water Permeability and Chloride Ion Diffusion in Portland Cement Mortars: Relationship to Sand Content and Critical Pore Diameter. *Cement and Concrete Research*, Vol. 25, No. 4, pp. 790-802.
5. Verbeck, G.J. and Helmuth, R.H. (1968) Structures and Physical Properties of Cement Paste. *Proceedings of the 5th International Symposium on the Chemistry of Cement.* pp. 1-32.
6. Kjellsen, K.O., Detwiler, R.J. and Gjørv, O.E. (1990) Backscattered Electron Imaging of Cement Pastes Hydrated at Different Temperatures. *Cement and Concrete Research*, Vol. 20, No. 2. pp. 308-311.
7. Kjellsen, K.O., Detwiler, R.J. and Gjørv, O.E. (1990) Pore Structure of Plain Cement Pastes Hydrated at Different Temperatures. *Cement and Concrete Research*, Vol. 20, No. 6. pp. 927-933.
8. Mehta, P.K. and Manmohan, D. (1980) Pore Size Distribution and Permeability of Hardened Cement Pastes. *7th International Congress on the Chemistry of Cement.*, Vol. III. pp. VII-1-5.
9. Detwiler, R.J., Kjellsen, K.O. and Gjørv, O.E. (1991) Resistance to Chloride Ion Intrusion of Concrete Cured at Different Temperatures. *ACI Materials Journal*, Vol. 88, No. 1. pp. 19-24.
10. Detwiler, R.J., Fapohunda, C.A. and Natale, J. (1994) Use of Supplementary Cementing Materials to Increase the Resistance to Chloride Diffusion of Concrete Cured at Elevated Temperatures. *ACI Materials Journal*, Vol. 91, No. 1. pp. 63-66.
11. Fapohunda, C.A. (1992) *Resistance to Chloride Intrusion of Blended Cement Concretes Cured at Elevated Temperatures.* M.A.Sc. thesis, University of Toronto Department of Civil Engineering.
12. Campbell, G.M. and Detwiler, R.J. (1993) Development of Mix Designs for Strength and Durability of Steam-Cured Concrete. *Concrete International*, Vol. 15, No. 7. pp. 37-39.
13. Cao Y.J. and Detwiler, R.J. (1995) Backscattered Electron Imaging of Cement Pastes Cured at Elevated Temperatures. *Cement and Concrete Research*, Vol. 25, No. 3. pp. 627-638.

11 OPTICAL MICROSCOPY AS A TOOL FOR DEVELOPMENT OF HIGH-PERFORMANCE CONCRETE

H.C. GRAN
Norwegian Building Research Institute, Oslo, Norway

Abstract
Optical microscopical analysis has been used in characterization of concrete in several countries during the last two to three decades. In several coutries, the technique has become an important method in evaluation of concrete quality/durability and rehabilitation as well as a governing parameter for concreting. Parameters as w/c-ratio, porosity, cracking, carbonization, homogeneity, AAR, types of aggregate may be determined. The technique has so far only covered concretes of higher w/c-ratios, introducing artifacts as cracking at lower w/c-ratios. This paper deals with the development and application of a modified microscopical technique, Fluorescent Liquid Replacement (FLR) technique which covers high-performance concretes down to w/(c+s)-ratios of 0.20. Round-robin tests and applications to frost testing, thermal cracking, shotcreting and evaluation of offshore concrete structures are covered.
Keywords: Cracks, Fluorescent Liquid Replacement (FLR), high strength concrete, microscopy, w/c-ratio.

1 Introduction

In the same way as many human activities, the use of concrete contributes in a negative way to our environment. Production of cement is highly energy consuming and releases large amounts of carbon dioxide to the athmosphere from both the rawmix and the burning of fuel. Man has a clear responsibility to minimize the negative environmental effects through more efficient use of cement. One obvious way to do this is to save energy and reduce the release of CO_2 indirectly through the production of concrete of better durability, preferably in combination with mix designs specifying lower contents of cement perhaps in combination with pozzolans like silica fume or fly ash. To

Integrated Design and Environmental Issues in Concrete Technology. Edited by K. Sakai. Published in 1996 by E & FN Spon, 2–6 Boundary Row, London, SE1 8HN, UK. ISBN 0 419 22180 8.

achieve this, a better understanding and knowledge about the microstructure and microstructural behaviour of concrete is essential. An important technique in the studies of microstructure is optical microscopy.

Optical microscopical analysis has been used in characterization of concrete and its microstructure in several countries for two to three decades. The technique is based on the microscopical examination of thin sections and plane sections. The technique has found application in areas as rehabilitation, evaluation of concrete quality/durability and research. Parameters as water/binder-ratio, cracking, volume ratios between concrete constituents, air void structure, homogeneity, carbonization, AAR and aggregate type may be determined.

This paper is not intended to give a presentation of the optical microcopical analysis in general, but a brief survey of the recent development and reported applications of a modified technique called Fluorescent Liquid Replacement (FLR) technique adapted to high-performance concretes. The traditional technique is described in detail elsewhere [1,2] and readers may refer to these publications for basic information on the subject.

In the microscopical analysis, some of the parameters like air content, cracking, w/c-ratio and homogeneity are determined by the use of ultra violet light. This is done by impregnating the concrete samples with a fluorescent dye and recording variations in fluorescense intensity resulting from differences in porosity.

Until now, preparation of concrete samples has been done by traditional impregnation with a fluorescent epoxy. With the advent and increased application of high strength concrete in various fields in the later years, the need for a modification of the technique to cover lower water cement ratios has become more and more prominent. As an example, reliable w/b-ratio determination in epoxy impregnated concretes is usually limited downwards to a w/c-ratio of 0.35. With addition of micro silica fume to the concrete even a w/c+s-ratio of 0.40 - 0.45 may cause problems. This limitation is caused by the macromolecular character and consequently high viscosity of resins and in some cases also hardeners.

To circumvent this problem and to increase penetration capability, experiments have been done with epoxy systems of considerably lower viscosities [3]. These experiments, however, have only been partly successful and involved health hazards caused by the comparatively high volatility of these epoxy systems. However, the development of a technique based on liquid replacement of capillary porewater with a solution of ethanol and a fluorescent solute has been more successful. The description and application of this technique forms the subject of the rest of this paper.

2 Fluorescent liquid replacement (FLR) technique

The Fluorescent Liquid Replacement technique [4] has been developed at the Norwegian Building Research Institute as a result of continuous work in this field for several years.

2.1 Preparation
Preparation of concrete using the FLR-technique is done by the use of 1.0 % solution of fluorescent dye in ethanol as impregnating medium. The most frequently used fluorescent dye is Hudson Yellow, Struers Epodye or similar with an absorbing

frequency of 515 nm and emitting frequency of 535 nm. Concrete samples are sawn to convenient size, ground and polished on one side before being immersed directly into the ethanol solution without drying. The samples are normally kept in the solution for four days. Experience shows that this period is sufficient for sufficient replacement of porewater by the fluorescent ethanol solution [4]. After slight drying in air at room temperature, the fluorescent dye is fixed by application of a thin layer of normal fluorescent epoxy (resin: BY 258 or similar; hardener: MY 2996). Excess epoxy on the surface is removed by very light grinding.

2.2 Comparison with traditional epoxy technique

The most significant effect of the FLR-technique is a large improvement of impregnation depth compared to the traditional epoxy technique. Table 1 below shows impregnation depths in mm reported by Gran [4] for different concretes using FLR-technique compared to epoxy impregnated concretes with and without silica fume. Impregnation depths were measured by sawing the samples perpendicular to their impregnated surfaces and inspecting the sawn surfaces in microscope using UV-light.

Table 1. Comparison of observed maximum impregnation depths, FLR impregnated [4] and epoxy impregnated concretes

water/binder-ratio	Content of silica fume (weight-%)	Impregnation depth (mm), epoxy	Impregnation depth (mm), FLR
0.20	10	-*	0.1
0.25	10	-*	1.5
0.40	10	-*	2.5
0.35	0	0.05	> 10

*) Below detection limit, 0.01 mm

The epoxy system did not show sufficient penetration capability for application to thin section production from high strength concrete. It was concluded that the pore-system in high strength concrete was so fine that only very small molecules with small intermolecular forces were allowed to enter. Very few epoxies are based on monomers that fulfil these requirements. The possible epoxies are restricted by high volatility and health hazards. Heating the epoxy and thus reducing its viscosity mainly through reduction of intermolecular interactions did not increase the impregnation depth. This was regarded as an indication of too large molecules rather than intermolecular forces.

Due to the high intensity exhibited by Hudson Yellow and similar fluorescent dyes, cracks of widths in the range from 1.0 μm and upwards are easily detected in the microscope using ultra violet light.

Comparison of traditional epoxy impregnation techniques with the FLR technique [5,6] showed that traditional techniques had a tendency to give artificial cracking in concretes in the lower w/b-ratio range. Concretes prepared by the FLR technique did not introduce similar cracking. Traditional epoxy impregnation requires drying at 50 °C and impregnation by vacuum due to the relatively high viscosity of epoxy systems. This treatment caused cracking. The FLR technique does not require drying before impregnation.

3 Applications

3.1 Water cement ratio determination

As mentioned before, the w/b-ratio determination is done by observing the fluorescense intensity in UV-light. During impregnation, the epoxy enters the capillary pore structure of cement paste and may be observed as a greenish glow by the application of ultra violet light and proper filtres. Measurement of the w/b-ratio is possible as a consequence of the proportionality between capillary porosity and the w/b-ratio. Porous cement pastes of high w/b-ratios are recognized by an intense glow whereas denser pastes of lower w/b-ratio turn out darker.

The FLR-technique has improved the range of water/binder-ratio determination which now extends down to w/b = 0.20 - 0.25, comprising high performance concretes containing up to 10 weight-% of silica fume. Gran [4] has plotted the fluorescense intensity vs. w/b-ratio as shown in Figure 1. The practical range for w/b-ratio determination with FLR technique is shown between 0.20 and 0.50. Fluorescense intensities measured at higher w/b-ratios than 0.50 are difficult to separate. The practical range of normal epoxy systems is usually w/b = 0.35 to w/b = 0.70 - 0.80. As regards w/b-ratio determination, the FLR technique must therefore be considered a supplement to the traditional technique.

Problems on poor reproducibility of w/b-ratio measurements have earlier been reported [7] for the traditional technique. To check the reproducibility of the FLR-technique a round robin test was carried out with participation of five different Nordic laboratories with NBI being responsible for project management [6]. In addition to obtain a measure of the reproducibility, a main purpose of the work was to get a general evaluation of the applicability and feasibility of the technique in different laboratories. The other participating laboratories were the Danish Technological Institute (TI), Denmark, Technical Research Center of Finland (VTT), Finland, Icelandic Building Research Institute (Rb), Iceland, Swedish National Testing Institute (SP), Sweden.

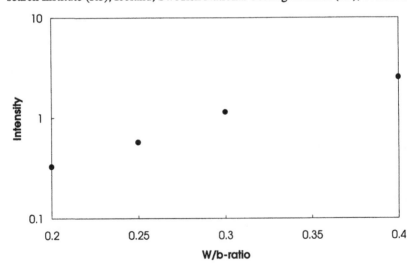

Fig. 1.   Fluorescence intensities plotted v.s w/b-ratio measured by Gran in plane sections [4].

High strength concretes were mixed with four different w/c+s-ratios, 0.20, 0.25, 0.30 and 0.40, using Norwegian Ordinary Portland Cement and containing 10 % of silica fume by weight of cement. The concretes were stored in water from the second day of curing after demoulding and attained 28 days compressive strengths of more than 100 MPa. A total of twenty 15 cm cubes to be used for thin section preparation were distributed between the laboratories. Maximum diameter of the aggregate, d_{max}, was 16 mm.

The test showed that various problems occurred partly due to inexperience with the new technique. Equipments and facilities for production of thin sections varied between the laboratories. The differences, however, did not influence on the final quality of the thin sections.

After preparation, all thin sections were circulated between the laboratories for water binder ratio evaluation. At the time of the test no reference material existed. Measurement of w/c-ratio was therefore limited to a relative comparison of the thin sections according to fluorescense intensity and the four known water binder ratios.

Of a total number of twenty-four thin sections which were evaluated in the five participating laboratories, totalling 120 separate measurements, only four samples were assigned incorrect w/c+s-ratios.

When necessary precautions are taken during preparation, the FLR-technique should therefore be considered for the most part to give sufficient reproducibility, and w/b-ratio determination in high strength concretes using FLR technique is now regarded as routine work.

3.2 Cracking

This part of the paper describes recent and present investigations on thermal cracking in concretes. As mentioned earlier, recent work has indicated that concretes of higher compressive strengths show a greater tendency of cracking when dried at different temperatures than normal concrete. In this investigation, the FLR-technique has been used to see the effect of slight drying on cracking in concrete. The work was initiated as a result of superficial micro cracking observed in a number of offshore concrete specimens that were exposed to natural weather conditions for a longer period.

Two drying procedures were selected. Drying at 50 °C for 12 hours and simulated sun exposure at different temperatures. A 500 W halogen lamp was used to simulate sun exposure. The tested concretes were 3 months old and consisted of one normal concrete with a w/c-ratio of 0.40 and one high strength concrete with a w/c+s-ratio of 0.30 with 10 % silica fume. The concrete specimens were 15 cm x 15 cm x 10 cm. The simulated sun exposure consisted of cycles of four hours sun exposure (at 50 - 70 °C) and four hours shadow (at 20 °C), numbering a total of four cycles. The results were compared to virgin samples (never dried). Cracking was observed as shown in Table 2.

Table 2. Extent of cracking due to thermal exposure of concrete specimens compared to virgin concrete specimens

W/c-ratio	Virgin	Dried at 50 °C	Simulated sun exposure at max. 69 °C
0.40	no	no	no
0.30 (10 % silica fume)	no	extensive	extensive

The concrete cast with w/c+s-ratio of 0.30 (10 % silica fume) showed an extensive cracking pattern both when dried slightly in laboratory at 50 °C and when exposed to cycled simulated sun. The normal concrete with w/c-ratio of 0.40 and virgin concrete showed no sign of cracking. The great difference is probably explained by different brittleness of the two concretes. The effect of this cracking on durability and strength has for the moment not been studied. Further work on the subject is in progress.

3.3 Frost cracking
FLR preparation was used by Bakke [5] and Jacobsen et al. [8] to investigate the correlation between cracking and ASTM C666 frost Durability Factor in concretes with different types of aggregate. Quantification of the amount of cracking was done by counting the number of cracks within a given area using a modified test developed by The Danish Institute of Technology [9]. Significant correlation was observed as shown in Figure 2.

3.4 Cracking in offshore concrete
Microscopical quality examination of offshore concrete using FLR technique revealed micro cracking in drilled cores from untreated concrete structures. Concretes coated with epoxy did not show similar cracking. The microscopical examination showed that the cracking was only superficial (to a depth of approximately 5 mm) and was thus considered to be of minor importance to durability.

3.5 Cracking in shotcrete
The FLR technique was used by Løken and Mørch [10] as part of a comprehensive investigation on the durability of shotcretes with compressive strengths up to 80 MPa. Microscopical analysis of plane sections showed extensive internal cracking with

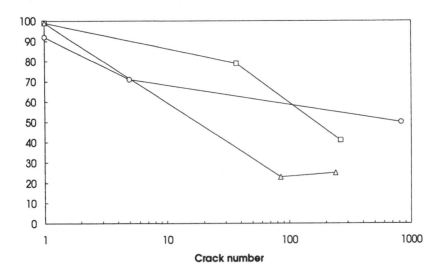

Fig. 2   Durability Factor vs. Crack number plotted for three different types of aggregate [5,8]

additions of sodium silicate accelerator at concentrations above 26 l/m^3. The analysis also showed that high levels of sodium silicate addition lead to inhomogeneous concrete with larger air voids and reduced binding between fiber reinforcement and cement paste. They concluded that this, in combination with cracking, accounted for poor frost resistance, high capillary absorption, high water permeability and reduced mechanical properties.

4   Concluding remarks

Optical microscopy in combination with FLR technique has been in use since 1992. During this period it has shown its potential in structural characterisation of high strength/high performance concretes. One of the most important aspects connected to the technique is that it has turned out to be an interesting tool for studies of cracking in concrete as the risk of cracking caused by the treatment during preparation is low.

5   References

1.   Sandström, M. (1988) *Microscopic estimation of the water-cement ratio in hardened concrete*. Borås, Swedish National Testing Institute, Report SP-AR 1988:43. 31 p. (in Swedish) NT Project No. 677-87.
2.   Mayfield, B. (1990) *The quantitative evaluation of the water/cement-ratio using fluorescence microscopy*, Magazine of Concrete Research, 42, No. 150.
3.   Lundevall, G., Gran, H.C. and Rønning, T.F. (1991), *Application of thin section technique on high strength concrete*, Report 5.10, Materials developement - High strength concrete.
4.   Gran, H. C. (1995) *Fluorescent Liquid Replacement (FLR) technique: A means of crack detection and w/c-ratio determination in high strength concrete*, Cement and Concrete Research, Vol 25, No 5.
5.   Bakke, J. A. (1993) Diploma thesis The Norwegian Institute of technology/The Norwegian Building Research Institute, (In Norwegian)
6.   Gran, H. C. (1994) *Determination of water cement ratio in high strength concrete by thin section technique*, NBI project report No. 147.
7.   Norwegian Building Research Institute, National Road Directory (1993) *Chloride resistance of concrete coastal bridges in Norway.*
8.   Jacobsen, S., Gran, H. C., Sellevold and E. J., Bakke, J. A. (1995) *High strength concrete - Freeze/thaw testing and cracking*, Accepted Cement and Concrete Research.
9.   Danish Institute of Technology (1987) Test method TI-B5 Denmark (In Danish).
10. Løken, J. and Mørch, A. (1994) *Durability of shotcrete, Part 1*, Diploma thesis The Norwegian Institute of technology/The Norwegian Building Research Institute, (In Norwegian)

12 INTEGRATIVE APPROACH TO OPTIMIZING THE DURABILITY PERFORMANCE OF REINFORCED CONCRETE RESIDENTIAL BUILDINGS

A. BENTUR
National Building Research Institute – Technion, Israel

Abstract
The present paper describes a case study of an investigation of the durability of the envelope of concrete residential buildings in Israel. It is an integrative work dealing with zoning according to environmental risks, materials research to evaluate and improve long term performance and techno-economical analysis of the various alternative solutions based on life cycle cost calculations. The intention of this presentation is to demonstrate the need and usefulness of such an integrative approach to deal with durability issues, and to emphasize that accepted durability performance should not always be addressed only from the point of view of the material's performance itself, which is judged only on the basis of its life span. In some applications, as the one considered in this paper, the system of superior durability is not necessarily the one giving the minimum life cycle cost. Thus research which is aimed at improving the life expectancy only, even if successful technically, may not yield the most efficient solution. It may turn out that the most efficient use of the limited resources for R&D is not necessarily in improvement and development of highly durable sophisticated materials, but rather in modification of existing materials and construction systems which can be seen to have the advantage when life cycle cost comparison is made.
Keywords: cladding, concrete buildings, durability, envelope, life cycle cost, life expectancy, rendering.

1 Introduction

Durability and repair issues in the field of concrete are usually concerned with special structures such as bridges and off-shore platforms. Yet, we should keep in mind that a large portion of the construction activity and costs are in the area of

Integrated Design and Environmental Issues in Concrete Technology. Edited by K. Sakai. Published in 1996 by E & FN Spon, 2–6 Boundary Row, London, SE1 8HN, UK. ISBN 0 419 22180 8.

residential housing, and although the damage caused by durability problems may not be as dramatic as in special structures (e.g. closure of a highway bridge for major repair), the cumulative cost to society may be larger.

The approach taken in the design for durability of special structures is in many instances less sensitive to direct costs as this is overshadowed by the economic damage that may occur if the structure ceases from being operational, even temporarily. As a result, much more sophisticated materials are justified in the construction of such special structures. In contrast with that, in residential housing the direct cost issue is expected to be more critical and therefore the approach and technologies to be used to ensure cost/effective durability performance may be different. Thus when dealing with rational design for durability, special attention should be given to the unique aspects of residential housing. In the present paper we will present a case study of an investigation based on integrated approach to durability of residential housing. It provides an overview of several coordinated research projects carried out in Israel.

2 Scope of problem and investigation

In Israel, like in many densely populated countries, residential housing consists largely of apartments in low to high rise buildings. They are usually made of a reinforced concrete structure with infill of concrete blocks, and an envelope of multi-layer cementitious rendering (stucco) with an external cementitious or synthetic coat, or alternatively, one layer of a cementitious coat which serves as a substrate to natural stone or ceramic tile cladding. Most of the durability problems and the required repairs in residential buildings are due to environmentally induced damage to the envelope, either corrosion of steel in the reinforced part of the structure and cracking and peeling of the external coat and cladding. Although the technologies for increasing the service life of these components are available, their rigid implementation in the codes is often met with resistance as the cost to the national economy may be prohibitive.

It was realized that a single solution imposed by the codes would not be an adequate answer to this problem, and the approach to be taken is to provide guidelines enabling to choose from a variety of technologies and technical means that would be most cost/effective for any particular situation. In the process of clarification of these issues it would also be possible to identify where modifications of existing technologies and the development of new ones can become attractive solutions. Thus, the approach taken was that of integrative studies, attempting to identify the following aspects: (i) influence of different environmental conditions on the service life so that the solution chosen would be adjusted to the local environmental conditions, (ii) analyzing the cost/effectiveness of various solutions for different environmental conditions, (iii) quantification of environmental conditions with respect to durability performance so they can serve as a basis for "zoning", and (iv) based on the cost effectiveness comparisons suggest modifications and new technologies with potential for improved technical and economic efficiency.

To achieve these goals the investigations conducted were quite diverse, including site studies, controlled evaluation of durability performance of materials in exposure sites and in accelerated tests, economic analysis based on the concepts of life cycle costs and laboratory research to explore the potential of improved technologies and materials.

3 Identification and quantification of environmentally corrosive zones

The major environmental differences in the various regions in Israel are not temperature (freezing hardly occurs even in mountainous regions) but rather the atmospheric conditions. They can be either due to presence of chlorides in areas in proximity to the seashore (result from sea water aerosol) and in areas where industrial and traffic pollution is extensive. Both types of environments are corrosive with respect to steel corrosion as well as cementitious and stone cladding.

Laboratory and site studies [1,2] showed that the influence of chlorides with regard to the severity of corrosion of steel in concrete and deterioration of cementitious rendering is similar, although the mechanisms are different. Evaluation of chloride penetration into cementitious systems, in conjunction with monitoring the chloride content in the air, enabled to provide guidelines for zoning, which is a function of the distance from the seashore and height of the structure [3] (Fig. 1). Such zoning and the methodology of its determination are similar to those reported in Japan [4]. Guidelines for each of these zones were suggested with regard to protection of steel in concrete (depth of concrete cover, quality of concrete, admixtures, etc.).

For performance of cladding a less detailed zoning was found suitable, defined as "corrosive" in zones within 1 km from the seashore and "non-corrosive" away from it. This zoning can provide more effective guidelines to the local designers, and

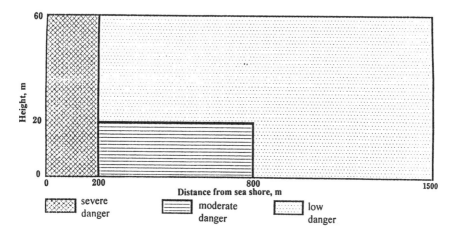

Figure. 1. Classification of corrosion risk zone as a function of distance from the sea shore and the height of the building, based on measurements of chloride content in the air and its penetration into concrete (adopted from Jaegermann et al [3]).

eliminates the difficulty of the practicing engineer in making decisions based on codes which provide qualitative classification of the environment into mild, moderate, severe and extreme. Yet, although the classification offered can be considered a useful forward step, it still leaves room for judgment based on micro-climate influences, and it still needs to be supplemented by guidelines regarding industrial and transportation corrosive zones. At present these are defined on the basis of qualitative assessment based on the presence of heavy industry or major intersections within about 2 km from the structure.

4 Durability performance and cost effectiveness of coating and cladding systems

Site studies of a large number of buildings were carried out [5,6]. The conditions of the envelope were graded on the basis of physical and visual ratings which were defined in terms of the nature and extent of damage (peeling, cracks, discoloration, etc.). Failure was defined when the weighted rating was less than 2 (on a scale of 1 to 5). It was found that the results could be presented and analyzed most consistently when the rate of failures of each category of cladding and coating was presented as a function of age and corrosiveness of the environment (Table 1). On that basis the life span of each kind of cladding/coating could be estimated, for corrosive and non-corrosive environments (Table 2).

It should be noted that in some of the categories which exhibited superior durability, the population of buildings surveyed was too young to enable assessment of the service life, and a minimum value is given. Attention should be drawn to the observations indicating that in this category failures were sometimes observed, usually within 10 years, but the inspection indicated that they were the result of poor application that did not comply with the codes. They were thus excluded from the data in Table 2.

Life cycle cost analysis based on the life expectancy data in Table 2 was carried out and the range of results obtained is shown in Figs. 2&3 [5,6]. The spread in the data for the stone cladding reflects also the range of life expectancy between 25 and 50 years. It can be clearly seen that for "non-corrosive" environment the most effective system is the cementitious stucco (Fig. 2), while for "corrosive" environment it is the synthetic stucco and the ceramic mosaic (Fig. 3). These results demonstrate that the approach of choosing solutions which are superior from performance point of

Table 1 Rate of failure in four types of coating/cladding systems of reinforced concrete structures (after Shohet et al. [6]).

Environment	Age	Coating/Cladding type			
		Cementitious	Synthetic	Ceramic	Stone
Non-Corrosive	≤10 years	7%	14%	29%	0%
	≥10 years	27%	43%	18%	20%
Corrosive	≤10 years	50%	18%	16%	0%
	≥10 years	89%	67%	25%	18%

Table 2 Life expectancy of four types of coating/cladding systems of reinforced concrete structures (after Shohet et al. [6])

Environment	Coating/Cladding Type			
	Cementitious	Synthetic	Ceramic	Stone
Non-Corrosive	10-15 yrs	12-15 yrs	>15 yrs	>25 yrs
Corrosive	<5 yrs	8-12 yrs	10-15 yrs	>25 yrs

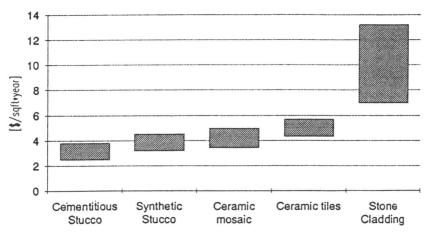

Figure 2. Range of life cycle cost of coating/cladding alternatives (building life span of 50 years) in non-corrosive environment (after Shohet et al [6]).

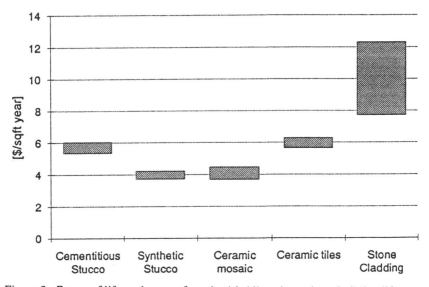

Figure 3. Range of life cycle cost of coating/cladding alternatives (building life span of 50 years) in corrosive environment (after Shohet et al [6]).

view (i.e. stone and ceramic cladding) is not necessarily the most cost effective. The data also suggests that from the point of view of R&D there is a greater potential for benefits if the cementitious stucco was further developed to improve its performance in the "corrosive" environment and extend its life span in the "non-corrosive" environment. The approaches taken are described in the next section.

5 Enhancement of the performance of cementitious coatings

The approach taken to enhance the performance of cementitious coatings incorporated two principles: Improve the material performance to enhance the life span and modify the composition so that its application on the structure is simplified, to reduce its cost. To achieve these goals laboratory study of small specimens and large panels was carried out and the conventional mixes which are essentially mixtures of Portland cement and lime were modified to consist of Portland cement mortar only, to which chemical admixtures or polymeric latexes were added [7,8]. Preliminary work for establishing criteria for long term performance, which included free and restrained shrinkage testing, concluded with two parameters upon which performance of different mixes was compared: a maximum value for free shrinkage, which is mandatory for minimum acceptance and a parameter which is the ratio flexural strength/free shrinkage. The latter can be used to compare between different compositions which complied with the maximum free shrinkage required criterion.

Optimal behavior was obtained with mixes containing certain types of acrylic and SBR latexes, as shown in Table 3. However, even with admixture which is added in a small amount, performance was considerably superior to the conventional mix. The comparison here is for a "conventional" system consisting of a cementitious base coat and a second leveling layer. The latter layer is the one where cracking usually occurs and this was the one modified with polymer [8].

Table 3. Effect of composition and the system of mortar for rendering on its performance (after Baum et al. [8]).

Mortar rendering composition	Free shrinkage (%)	Flexural strength (MPa)	Flexural strength to shrinkage ratio	Bond strength (MPa)		
				With base coat	Without base coat	
					Without primer	With primer
Conventional (Cement-Lime)	0.070	0.6	8.6	0.42	--	--
Cement + Workability and air entraining admixture (0.2%)	0.030	2.1	70	0.41	0.05	0.44
Cement + Acrylic latex (10%	0.014	4.8	343	0.47	0.54	0.44
Cement + SBR latex (10%)	0.008	3.6	450	0.50	0.21	0.53

As the cost of materials is increased with this kind of improvement, an additional step was taken to lower the cost, by eliminating the base layer which is usually the intermediate layer required to provide bond to the substrate (without it the bond is low, being about 0.05 MPa). The strategy taken was to replace the base layer which involves tedious application, with a thin latex primer that can be easily brushed or sprayed on the substrate. Proper choice of the ingredients enabled to obtain a system where the bonding is as good as when the base layer is present (Table 3).

6 Summary and conclusions

The present paper dealt with a case study of the investigation of the durability of the envelope of concrete residential buildings in Israel. Thus, some of the conclusions are relevant only to local conditions. However, the intention of this presentation was to demonstrate the need and usefulness of integrative approach to deal with durability issues, and emphasize that accepted durability performance should not always be addressed only from the point of view of the materials performance itself, which is judged only on the basis of its life span. In some applications, as the one considered in this paper, the system of superior durability is not necessarily the one giving the minimum life cycle cost. Thus research which is aimed at improving the life expectancy only, even if successful technically, may not yield the most efficient solution.

Thus the rational way to deal with such issues should involve several types of integrative studies to provide the following information and guidelines:

1. Evaluation of local environmental conditions with respect to durability performance of the component in question and setting "geographic zoning" accordingly.
2. Evaluation of the life span of different products in each of these zones, and determining the most cost-effective solution, i.e. the one of lowest life cycle cost. The optimal solution may be different for the different zones.
3. Decisions on spending R&D resources for improvement of durability performance should take into account existing data on life cycle costing. It may turn out that the most efficient use of the limited resources for R&D is not necessarily in improvement and development of highly durable sophisticated materials, but rather in modification of existing materials and construction systems which can be seen to have advantage when life cycle cost comparison is made.

7 References

1. I. Soroka and D. Carmel. "Durability of Rendering Mortar in Marine Environment", Research Report, National Building Research Institute, Haifa, Israel, 1990 (in Hebrew).
2. C. Jaegermann and D. Carmel. "Corrosion of Reinforcing Steel in Concrete in Marine Environment in Israel", Research Report, National Building Research Institute, Haifa, Israel, 1991 (in Hebrew).

3. C. Jaegermann, H. Baum and D. Carmel. "Influence of distance from Seashore on the Chloride Content Penetrated into Concrete", Research Report, National Building Research Institute, Haifa, Israel, 1990 (in Hebrew).

4. N. Kashino. "Some Aspects of the Classification of Pollution by Sea Water Aerosol for Reinforced Concrete Building", Ocean Space Utilization-85, Proc. Int. Symp., Tokyo, June, 1985, Vol. 2, pp. 447-454.

5. I.M. Shohet. "Comparison Between Exterior Cladding Systems in Buildings", Research Report, National Building Research Institute, Haifa, Israel, 1995 (in Hebrew).

6. I.M. Shohet, A. Laufer and A. Bentur. "Exterior Cladding Methods: Techno-Economic Analysis", submitted for publication.

7. I. Soroka and H. Baum. "Factors Affecting the Cracking of Exterior Rendering Mortar", Research Report, National Building Research Institute, Haifa, Israel, 1991 (in Hebrew).

8. H. Baum, I. Soroka and A. Bentur. "Improved Mortar for Rendering: Admixtures", Research Report, National Building Research Institute, Haifa, Israel, 1994 (in Hebrew).

13 DETERIORATION MECHANISM OF GLASS FIBRE REINFORCED CONCRETE AND PREDICTION OF STRENGTH REDUCTION

T. UOMOTO
Institute of Industrial Science, University of Tokyo, Tokyo, Japan
F. KATSUKI
Department of Civil Works Technology, Tokyu Construction Co., Tokyo, Japan

Abstract
To utilise GRC as a structural material, the deterioration mechanism of GRC must be known. This paper describes how glass fibres are deteriorated due to alkali, and explains why the GRC loses its strength after a long period of time. The paper also proposes how to estimate the strength reduction of GRC. The estimated results match well with experimental data of others.
Keywords: GRC, glass prediction, modulus of rupture

1   Introduction

Glass fibre reinforced cement and concrete(GRC)has been used in many countries for a long period of time. There are several advantages when compared with steel fibres: fibre do not corrode due to salt, the appearance is better in case of GRC compared to SFRC, etc.

Due to the reaction between glass and the alkalis in cement, one big problem for GRC is that strength reduction may occur over a period of time. To prevent the reaction, several types of alkali-resistant glass fibres have been developed. If we can predict the strength change of GRC by numerical methods, it can be used not only to remote GRC as a structural   material   but   also to evaluate the degree of alkali-resistance of glass fibres.

Considering   these, this   paper   clarifies the deterioration mechanism of   both glass fibres and GRC quantitatively. These mechanism of GRC are validated by comparing the calculated results of modulus of rupture with experimental results, a prediction method is also proposed in this paper.

2   Experimental procedure

2.1   Accelerated test of fibres
Glass fibres used for experiment were T-glass fibres, and Table 1 shows the material properties. Table 2 shows the condition of   accelerated test of glass fibres which were immersed

Integrated Design and Environmental Issues in Concrete Technology. Edited by K. Sakai. Published in 1996 by E & FN Spon, 2–6 Boundary Row, London, SE1 8HN, UK. ISBN 0 419 22180 8.

in aqueous NaOH.

In the experiment, concentration of aqueous NaOH varied from 0.05 to 2.0 mol/l, temperature ranged from 20 to 60℃,   and curing time varied from 24 to 676 hours. After these, fibres were washed with distilled water and dried for 24 hours in a desiccator.

Table 1.   Chemical compositions and properties of T-glass fibre

Diameter ; 12.8 μm			Tensile strength ; 2510 MPa			
ig.loss	SiO2	Al2O3	Fe2O3	CaO	MgO	Na2O
0.8	65.1	23.9	.108	.296	9.74	.084

Table 2. Conditions of accelerated test

Alkaline conc. (mol/l)	0.05	0.5	1.0	2.0
Tempreature (℃)	40	40	20, 40, 60	40

2.2   Tensile test

Figure 1 shows a typical specimen of glass fibre used for a tensile test. The fibres after accelerated test were separated into mono-filaments and then each fibre was bonded on to testing papers in   accordance with JIS-R-7601. Tensile tests were carried out using displacement controlled (5kgf) machine at room temperature (Temperture:20±3℃). Number of specimens varied from 50 to 60 per case maintaining the cross-head speed at 0.5 mm/min.

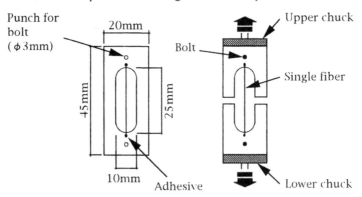

Fig.1.   Specimen used for tensile tests

3   Mechanism of deterioration

Photo 1 shows the conditions of T-glass fibres observed by SEM with exposure times of 0, 7 and 14 days. Here, alkaline   concentration was 1.0mol/l, and the temperature was 40℃. It shows clearly that non-immersed fibre has very surface compared to immersed fibres, and it reveals   that   chemical   reaction has taken place on the surface of immersed fibres. From   the photographs of T-glass fibres immersed for 7 and 14 days, the deteriorated portion increased proportionally from the   surface   of   fibres inward with time. Photo 2 shows an enlarged view of the part corroded by alkali. It is found that these products are porous. Therefore, it can be assumed that tensile strength loss of T glass fibres is caused by the decrease of the original cross sectional area due to the alkali attack.

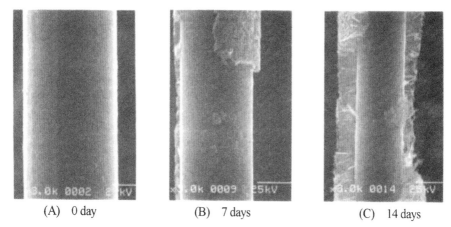

| (A)   0 day | (B)   7 days | (C)   14 days |

Photo.1.   The conditions of T-glass fibers (SEM)

4   Model to predict the progress of deterioration

4.1   Deterioration model due to alkali penetration

Deterioration of   T glass fibre due to alkali was modelled on the basis of the result (photo1) in SEM observation. The authors considered that the progress of reaction accompanied by   diffusion of   alkali   is from the surface to the inner layer of T glass fibre. Therefore   it can   be   assumed that the rate of alkali-silica reaction on T glass fibre is determine by the rate of diffusion of hydroxide ion from the surface to the inside of the fibre.

Figure 2 shows a model to predict the progress of deterioration. In this model, the progress of deterioration can be simulated; the thickness of non-reacted layer can be calculated at a certain time based on the following equation.

$$\frac{\partial C}{\partial t} = D \cdot \left(\frac{\partial^2 C}{\partial t^2} + \frac{1}{r} \cdot \frac{\partial C}{\partial t} \right) + F(r,t;C) \tag{1}$$

where r, C, t indicate the radius   of   non-reacted   layer (mm), the alkali concentration (mol/l), and curing time (hour) respectively. D is the diffusion coefficient (cm^2/hour) of alkali in fibre.

The function F(r,t;C), in equation (1), indicates alkaline consumption due to reaction between   alkali   and   silica   in   glass   fibre.   Actually, the quantity of alkali which   silica can   consume   per   unit   volume   limits   the value of the product by the quantity   of silica per unit volume   and   the rate of alkali-silica reaction on T glass fibre. This value is defined as "Quantity of Alkali Effective Consumption (QAEC)".

Therefore, in the analysis, when the quantity of alkali accumulated to each section of fibre exceeded this QAEC, the reacted layer is formed. The solution to equation (1) was carried out using Finite Differential Method.

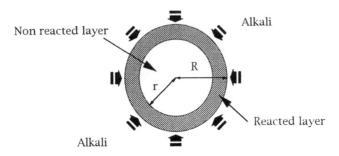

Fig.2.   Analytical model to predict the process of degradation

4.2   Diffusion coefficient and QAEC of T glass fibre

The quantity of alkali-silica reaction   calculated in equation (1) can be evaluated by QAEC and diffusion coefficient which   influences   penetration   speed. The authors have calculated the diffusion coefficient at 40°C using Fick's first law ($D=1.0\times10^{-8}$ mm^2/hour), and the rate of alkali-silica reaction of fibres has been determined based on that of quartz glass. Hence, QAEC of T glass fibre is 1.3 mol/l.

5   Results of analysis and discussion

5.1 The influence of temperature change on the rate of reaction

Based on experiments, it is found that the effect of temperature, as the variation of diffusion coefficient of alkali corresponding to temperature change, can be expressed by the following Arrhenius's equation.

$$\log_{10} D = \left(-\frac{E}{R\cdot T}\right) + \log_{10} F \tag{2}$$

where, F, E and T indicates frequency   factor, activation energy and absolute temperature respectively. Therefore, by using this equation, suitability of diffusion coeficient was examined by calculation from the results of tension test at each condition of temperature.

The formula to calculate immersed load carrying capacity of glass fibre at a certain age is performed under the following assumptions.

   (1)   Immersed fibre strength is decided using the thickness of non-reacted portion which is calculated by equation(1); namely, tensile strength of the reacted portion is 0.
   (2)   The tensile strength of non-reacted portion is the same as that of virgin fibres.

From the assumption (1) and (2), the following equation can be obtained.

$$\sigma_0 = \frac{P_0}{S_0} = \frac{P_t}{S_t}$$
$$\therefore P_t = S_t \cdot \sigma_0 \tag{3}$$

where σ_0 indicate   tensile   strength   before   immersing (MBa). P_0 and P_t indicate failure load before immersing and at a certain age t (kN). S_0 and S_t indicate sectional areas of the glass fibre before and after the immersion. Therefore, tensile strength of immersed glass fibre σ_i(Mpa) can be calculated as follows.

$$\sigma_t = \frac{P_t}{S_0}$$

$$= \sigma_0 \cdot \frac{S_t}{S_0}$$

$$= \sigma_0 \cdot \frac{r^2}{R_0^2} \tag{4}$$

The strength loss of glass fibres in accelerated test at 20, 40 and 60°C is shown in figure 3. Table 3 shows the diffusion coefficients which was calculated by fitting the experimental data of figure 3. These diffusion coefficients are illustrated by using Arrhenius's equation (2) in figure 4. As the relation between log D and 1/T is linear, and activation energy calculated from regression analysis of figure 4 is 25.6 kcal/mol, the validity of equation (2) is justified.

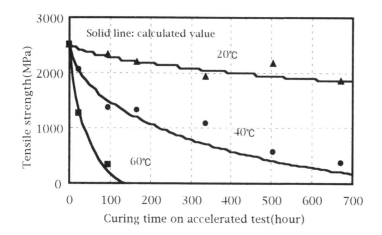

Fig.3. The strength loss of accelerated tests at 20, 40 and 60°C with 1.0mol/l aqueous NaOH

Table 3. Calculated diffusion coefficient($\times 10^{-9}$ mm^2/hrs)

20°C	40°C	60°C
1.02	10.00	18.14

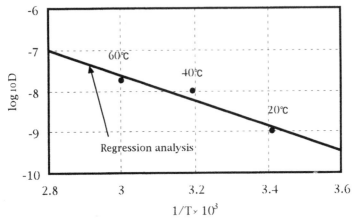

Fig.4.   Relation between log D and 1/T(Arrhenius's plot)

5.2   Influence of alkaline concentration on the reaction speed

The change of diffusion coefficient due to alkaline concentration was also investigated. Figure 5 shows the strength loss due to alkaline concentration of 0.05, 0.5 and 2.0 mol/l. Here, temperature   was   kept at 40°C. Also in this figure, the values were calculated by using the diffusion coefficient for 40°C with different concentration of aqueous NaOH. As shown, even with different concentrations of NaOH and but the same diffusion coefficient at 40°C, the result of simulation correspond well to experimental results. That is to say, that the alkaline concentration does not influence the diffusion coefficient.

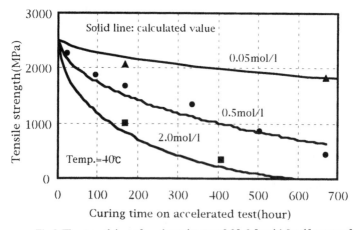

Fig.5. The strength loss of accelerated tests at 0.05, 0.5 and 1.0mol/l aqueous NaOH at 40°C

5.3   Prediction of deterioration of glass fibre under actual   environmental condition

Table 4 shows the results of predicted strength loss of glass fibres under 2 conditions. One is calculated with an alkaline concentration of 1.0 mol/l   (pH=14) and a temperture of 40°C. The other, supposing exposure under actual environmental condition, was caluculated with an alkaline concentration of 0.05 mol/l (pH=12.7) and a temperture of 20°C. As an example, for a strength loss of 92%, it takes about 21 years under actual environment and only 28 days in an accelerated test.

Table 4.   Prediction of deterioration of T-glass fibre under actual environmental condition

Curing time on accelerated test (day)	Exposure time under actual environmental condition (year)	The rate of tensile strength loss (%)
1	0.7	19
4	3.0	41
7	5.2	52
14	11.3	72
21	15.9	83
28	20.9	92
42	33.8	100

Temp.=40℃	Temp.=20℃
pH=14.0 (1.0mol/l)	pH=12.7 (0.05mol/l)

6   Evaluation of durability of GRC

When GRC is exposed under wet conditions for a long period of time, mechanical properties such as flexural strength decreases with time. In general, it has been shown that two factors could be the cause of this decrease. One is the poor alkali resistance of glass fibres, and another is the change in the micro structure caused by deposition $Ca(OH)_2$ between glass fibres and the matrix. In this paper, it is assumed that the strength of GRC is decreased by deterioration of glass fibres due to alkali, and that flexural strength loss of GRC can be predicted by using the rate of tensile strength loss of glass fibre as in table 4.

6.1   Bending theory of GRC

It is assumed that distribution of tensile stress is rectangular and that does not fail first. Figure 6 (a) shows stress distribution of the GRC section.

From the balance of tensile force "C" and compressive force "T" in the figure,

$$T = \sigma_t \cdot (d - x) = \frac{1}{2} \cdot \sigma_c \cdot x = C$$

where, $x = \dfrac{2 \cdot d}{\alpha + 2}$ when the equation $\sigma_c = \alpha \cdot \sigma_t$, is assumed.

Flexural moment "M" is given by

$$M = \frac{b}{2} \cdot \sigma_c \cdot x \cdot (\frac{2}{3} \cdot x + \frac{1}{2} \cdot (d - x)).$$

Therefore, flexural strength σ_B is given by the following equation.

$$\sigma_B = \frac{M}{I} \cdot \frac{d}{2}$$

$$= \frac{\sigma_t \cdot (3 \cdot \alpha + 8) \cdot \alpha}{(\alpha + 2)^2}$$

$$= \beta \cdot \sigma_t$$

(5)

where $\beta = \dfrac{(3\alpha + 8)\alpha}{(\alpha + 2)^2}$

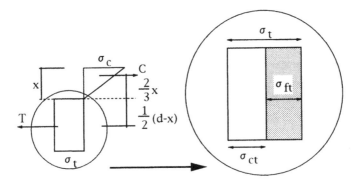

Fig.6.   Bending theory of GRC

6.2   Prediction of flexural strength of GRC

In this paper, it is assumed that the sum of both tensile strength of matrix "σ_{ct}" and tensile strength of glass fibres "σ_{ft}" is equal to the tensile strength "σ_t" of GRC as shown figure 6(b).

Accordingly σ_t is given by the following equation.

$$\sigma_t = \sigma_{ct} + \sigma_{ft}$$

By using the rate of tensile strength loss of glass fibre at certain age "$\gamma(t)$", the tensile strength loss of GRC at certain age "$\sigma_t(t)$" is given by

$$\sigma_t(t) = \sigma_{ct} + \sigma_{ft} \cdot \gamma(t)$$

(6)

Substituting equation (6) into equation (5) gives,

$$\sigma_B(t) = \left(\sigma_{ct} + \sigma_{ft} \cdot \gamma(t)\right) \cdot \beta$$

(7)

Figure 7 shows the result of calculation according to equation 7.   In the figure, the results of bending test of GRC, of which deterioration was accelerated, carried out by B.A.Proctor, is also plotted. It is noticed that experimental data in this figure were changed to those at a standard temperature by using Arrhenius's equation. Table5 shows the properties of strength of GRC without accelerated curing. Coefficients for calculation were chosen using values in the table 5.

$$\sigma_c = 60MPa, \quad \sigma_t = 17MPa, \quad \alpha = \frac{\sigma_c}{\sigma_t} = 3.53, \quad \beta = 2.15, \quad \sigma_a = \frac{1}{10}\cdot\sigma_c = 6.0MPa$$

$\sigma_B(t)$ in figure 7 was calculated by using the rate of tensile strength loss of T-glass fibre in table 4. It was clarified that calculated flexural strength coinside well with experimental data and that after 34 years the value was kept to be constant with the value of 12 MPa. This phenomenon shows that glass fibres can not contribute to flexural strength of GRC after 34 years, and flexural strength of GRC finally becomes equal to flexural strength of the matrix itself. Although adhesive property between matrix and glass fibbers, and the diminution of glass fibres are not considered in this analysis, it is possible to predict flexural strength of GRC by the proposed method over a long time

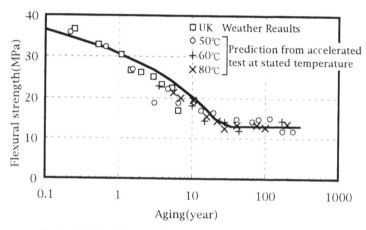

Fig. 7.   Relation between flexural strength and aging

Table 5.   Properties of GRC

PROCESS		SPRAY DEWATERED
Dry bulk density	t/m³	2.0 - 2.05
Compressive strength	MPa	60 - 100
Modulus	GPa	13 - 25
Impact strength	Nmm/mm²	11 - 25
Poisson's Raitio		0.24
Bending;		
LOP	MPa	9 - 13
MOR	MPa	28 - 42
Direct Tension;		
BOP	MPa	7 - 9
Strain to failure	%	10 - 17
Shear;		0.7 - 1.3
In-plane	MPa	10 - 17
Interlaminar	MPa	3 - 5

7 Conclusions

The following conclusions can be made:

1) Glass fibre reduces in strength in an alkaline environment due to a reduction in its diameter from the surface by the reaction between fibre and alkaline solution.
2) Both concentration of alkaline solution and temperature effect the reaction rate.
3) Tensile strength of glass fibres in alkaline solution can be predicted quantitatively by the proposed method using diffusion theory.
4) Modulus of rupture of GRC can be predicted by using bending theory and strength of glass fibres. The estimated result coincide well with experimental data.

8 References

1. Futoshi Katsuki and Taketo Uomoto (1994), Fundamental Study on Quantitative Evaluation of the Deterioration of Glass Fibre by Alkaline Solution. Proceedings of the Japan Concrete Institute, Vol.16, No.1. pp.1179-84
2. Continuous Fibre Research Committee (1992), Application of Continuous Fibre Reinforcing Materials for Concrete Structures. Concrete Library No.72 , pp.29-39
3. Proctor B.A. (1987), The Development and Performance of Alkali Resistant Glass Fibre for Cement Reinforcement . Proceedings of the International Symposium on Fibre Reinforced Concrete, December 16-19, pp.3-21
4. Hodhod H. and Uomoto T.(1992), Evaluation of Tensile Strength of FRP with Aligned Continuous Fibre. ADVANCED COMPOSITE IN BRIDGES AND STRUCTURES, pp.63-72
5. Uomoto T., Nishimura T. and Miyazaki F.(1993) Properties of FRP Tendons for Prestress (6). SEISAN-KENKYU, Vol.45, No. 5,pp.53-55
6. Litherland K.L and Proctor B.A (1985) The effect of matrix Formulation , Fibre Concrete and Fibre Composition on the Durability of Glass Fibre Reinforced Concrete. Proc. of Durability of Glass Fibre Reinforced Concrete , Sympo. , P.C.I, pp.124-35
7. P. Soroushian and A.Tlili (1993) Durability Characteristics of Polymer-Modified Glass Fibre Reinforced Concrete. ACI Materials Journal, V. 90, No. 1,   pp.40-9
8. Igarashi and Kawamura (1992) Microscopic Structure and Durability of Glass Fibre Reinforced Concrete. Proceeding of JSCE , pp.215-24

14 DURABILITY OF CONCRETE EXPOSED IN MARINE ENVIRONMENT FOR 25 YEARS

T. FUKUTE
Port and Harbour Research Institute, Ministry of Transport, Yokosuka, Japan

Abstract

In this study, long term durability of marine concrete is investigated. This study was started in 1970 at Port and Harbour Research Institute, Ministry of Transport, Japan. Many specimens with several different factors, such as type of cement, mixing water and cover thickness of reinforcing steel in concrete, were fabricated and exposed under marine environment. During 20 years exposure, several properties of concrete and reinforcing steel have been tested both chemically and physically. From the test results, several important conclusions were derived. In this paper, some findings from a series of tests are presented.

Keywords: Durability, exposure test, compressive strength, mix potential, anodic polarization, salt content, diffusion coefficient of oxygen.

1 Introduction

The 20th day of July has been called as "Marines Day" in Japan which is surrounded by ocean all around. Japanese government recently decided it to be the fifteenth Japanese national holiday after 1996. The role of marine and marine facilities is thought to become more important. On the other hand, in terms of ecology or environments, facilities and structures which consume high amount of energy and materials for their construction should be utilized for long time even under severe environmental condition such as marine environments.

Under these circumstances, achievement of highly durable marine concrete structures is one of the most important technical subjects to be solved urgently in our generation.

Integrated Design and Environmental Issues in Concrete Technology. Edited by K. Sakai. Published in 1996 by E & FN Spon, 2–6 Boundary Row, London, SE1 8HN, UK. ISBN 0 419 22180 8.

Port and Harbour Research Institute (PHRI) started marine exposure tests of various concrete test specimens in Kurihama, suburbs of Tokyo in 1970. Since that time, the effects of type of cement as well as mixing water on the long–term strength characteristics of concrete, chemical composition of concrete, and the corrosion characteristics of reinforcement have been surveyed. This research is on a physical, chemical and electrochemical consideration of the test results obtained during 20 years.

2 Test specimens and test items

2.1 Materials used

Seven types of cement, which are ordinary portland cement (symbol: N0, N2), high early strength portland cement (HO), moderate heat portland cement (MO), blast furnace slag cement B–class in JIS R 5211 (B0, B2) and aluminous cement (AL), were used. In the N2 and B2, the quantity of gypsum phosphate of 2% by weight of sulfur trioxide (SO3) was added to the N0 and B0 respectively[1]. Tables 1 to 3 show the chemical composition and physical test results of the cement.

Table 1 Chemical components of cement

cement	igloss (%)	insol. (%)	Al_2O_3 (%)	Fe_2O_3 (%)	S_iO_2 (%)	CaO (%)	MgO (%)	SO_3 (%)	Na_2O (%)	K_2O (%)	S (%)	fcao (%)
N0	0.8	0.3	5.5	3.1	21.7	64.7	1.3	2.0	0.30	0.57	—	0.9
N2	1.7	0.3	5.2	2.9	20.7	63.3	1.2	3.9	0.27	0.56	—	0.9
HO	1.1	0.5	5.2	2.7	20.6	65.3	1.2	2.5	0.37	0.54	—	0.9
MO	0.8	0.2	4.5	3.8	23.5	63.4	1.2	1.7	0.28	0.48	—	0.5
B0	0.4	0.7	9.7	2.1	25.7	54.1	3.2	2.4	0.46	0.51	0.4	0.3
B2	1.3	0.6	9.3	2.0	24.4	53.2	3.0	4.3	0.44	0.48	0.4	0.4

Table 2 Physical properties of cement (N0, N2, HO, MO, B0, B2)

Type of Cement	Specific gravity	Specific surface area (cm^2/g)	Setting time W/C (%)	Initial (h-m)	Final (h-m)	Flow (mm)	Compressive strength(N/mm^2) 3d.	7d.	28d.
N0	3.14	3180	27.5	2-28	3-28	250	12.8	22.1	41.4
N2	3.12	3570	27.8	2-50	3-50	249	15.6	22.6	37.2
HO	3.13	4290	29.0	2-23	3-26	259	22.1	33.1	43.8
MO	3.19	3260	25.5	3-02	4-27	255	8.8	14.3	31.1
B0	3.04	3850	28.8	3-34	5-10	255	9.5	14.9	33.9
B2	3.00	4100	29.0	3-54	5-52	253	9.3	14.9	31.9

Table 3   Physical properties of cement (AL)

Chemical component (%)			Specific gravity	Specific surface area (cm²/g)	Flexural strength (1 day) (N/mm²)	Compressive strength (3 days) (N/mm²)
Al₂0₃	Fe₂0₃	CaO				
60.6	2.5	32.5	2.98	4230	6.6	45.9

Tap water (W) and seawater (S) were used as mixing water of concrete.

As for coarse and fine aggregate, river gravel (maximum size: 25 mm) and river sand produced at the Sagami River, Kanagawa Prefecture were used.

All the concrete was mixed with water reducing agent as admixture; for (AL, W), air entraining agents were also used. As for the reinforcement, ϕ 9mm round bars were used in a brilliant condition; the black layer was removed.

2.2 Test specimens

For a series of tests, cylindrical test specimens (ϕ 15 × 30 cm) were fabricated with the concrete presented in Table 4.

Table 4   Specified mix proportions of concrete

Cement	Water	G_{max} (cm)	Slump (cm)	Air (%)	W/C (%)	S/a (%)	Unit quantity (kg/m3)					
							W	C	S	G	WE (1)	AE (cc)
N0	W	25	66	34	52.7	37.0	153	290	740	1261	2.9	—
	S	25	56	32	53.4	36.0	155	290	718	1277	2.9	—
N2	W	25	64	40	54.5	37.0	158	290	734	1251	2.9	—
	S	25	51	34	55.2	36.0	160	290	713	1270	2.9	—
H0	W	25	37	39	53.1	37.0	154	290	738	1258	2.9	—
	S	25	57	31	55.2	36.0	160	290	711	1263	2.9	—
M0	W	25	65	48	52.4	37.0	152	290	742	1264	2.9	—
	S	25	46	40	53.1	36.0	154	290	720	1280	2.9	—
B0	W	25	35	30	52.4	37.0	152	290	738	1258	2.9	—
	S	25	40	38	53.1	36.0	154	290	716	1274	2.9	—
B2	W	25	42	38	54.8	37.0	159	290	729	1242	2.9	—
	S	25	47	41	55.5	36.0	161	290	708	1258	2.9	—
AL	W	25	63	31	52.1	37.0	151	290	737	1256	2.9	20
	S	25	51	35	52.8	36.0	153	290	716	1272	2.9	—

* Slump and Air content are measured values.

After one week of curing in water, they were moved to exposure site. For the exposure test, a tidal pool in PHRI which can simulate tidal movement has been used. The tidal pool is supplied with seawater from Kurihama Bay and drained by the pump twice a day.

2.3 Main test items
- · Compressive Strength of Concrete
- · X-ray Diffraction Analysis
- · Corroded Area of Reinforcement in Concrete
- · Mix Potential of the Reinforcement in Concrete
- · Anodic Polarization of Reinforcement in Concrete [2]
- · Salt Content in Concrete [3]
- · Diffusion Coefficient of Oxygen in Concrete [4]

3 Test results and discussion

3.1 Compressive strength of concrete

Fig. 1 shows the time-dependent change of the compressive strength f'c of seawater mixed concrete. Fig. 2 shows the time-dependent change in the strength ratio, which is given by dividing the strength at each stage f'c by the initial strength f'co (hereinafter, called initial strength ratio) of sea water mixed concrete. Fig. 3 shows the change of the initial strength ratio of tap water mixed concrete. As for the relation between the aging and strength, it increases up to 5 years in most test specimens; however, after 5 years of aging the strength has a tendency to fall gradually; at the age of 20 years the strength is less than the initial strength. Such tendency of strength of concrete exposed to seawater has been reported often [5,6,7].

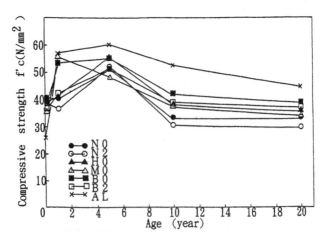

Fig. 1 Strength of concrete mixed with seawater

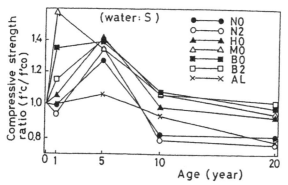

Fig. 2 Compressive strength ratio (mixed
with sea water)

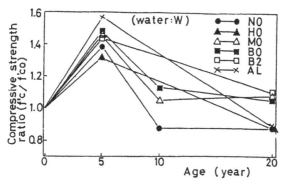

Fig. 3 Compressive strength ratio
(mixed with tap water)

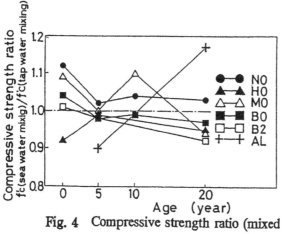

Fig. 4 Compressive strength ratio (mixed
with seawater v s tap water)

Fig. 4 shows the time–dependent change of the compressive strength ratio of concrete using seawater and tap water as mixing water (hereinafter, called the seawater/tap water strength ratio). Since concrete using seawater has a little more early strength than that using tap water, the compressive strength ratio of most test specimens at 28 days aging was more than 1.0. However, later, the compressive strength ratio approaches 1.0, and at 20 years of aging it became 0.92 to 1.03.
Thus,it is considered that the use of seawater for mixing is independent of the compressive strength of the long–term aging of concrete. This was also reported by Gjorv[3].
Table 5 shows the results of the X–ray diffraction analysis of the concrete at 20 years of aging. From Table 5, the following may be said.
① The ettringite was recognized in all samples; there is hardly a difference among the generated quantity by the types of cement and mixing water.
② The Friedel s salt was recognized in all samples. However, it was not detected in the center of the specimens mixed with blast furnace slag cement and tap water. It may be suggested that the rate of penetration of a chloride ion in this mixture is slow.

Table 5 X–ray analysis(N0, N2, H0, M0, B0, B2)

Cement	Water	Depth(cm)	Ettringite	Friedel's salt	CaCO3	Ca(OH)2
N 0	W	1	+	++	+	
		3	+	++	+	
		5	+	++	+	
		7.5	+	++	+	
	S	1	+	++	+	
		7.5	+	++	+	+
N 2	W	1	+	++	+	
		7.5	+	++	+	
	S	1	+	++	+	
		7.5	+	+		+
H 0	W	1	+	++	+	
		7.5	+	++	++	
	S	1	+	++	+	
		7.5	+	++	+	+
M 0	W	1	+	++	+	
		7.5	+	+	+	
	S	1	+	++	+	
		7.5	+	++	+	+
B 0	W	1	+	++	+	
		3	+	+	++	
		5	+	+	+	+
		7.5	+		+	+
	S	1	+	++	+	
		7.5	+	+	+	+
B 2	W	1	+	++	+	
		7.5	+	+	+	+
	S	1	+	+	+	
		7.5	+	+	+	+

+ : CPS (Counts per second) is less than 500
++ : CPS (Counts per second) is ranged from 500 to 2000

③ Ca(OH)2 was not detected on the surface of all the test specimens; particularly concrete mixed with portland cement and tap water Ca(OH)2 was not detected even in the center of the test specimens. It was found that Ca(OH)2 remains near the center of the specimens of the concrete with blast furnace slag cement group and the concrete mixed with seawater. These results suggest the following; the elusion of Ca(OH)2 from the concrete surface can be reduced by using the blast furnace slag cement or seawater as mixing water, but it can not be avoided under a marine environment over 20 years.

④ CaCO3 generated by the reaction of Ca(OH)2 and CO2 was recognized in all samples.

From these results, the strength lowering of the portland cement and blast furnace slag cement group under marine environment, particularly the portland cement group, are supposed to be caused by both of the reaction of Ca(OH)2 in concrete with penetrating chloride ions and the expansion due to generated ettringite formation. There is no sign that seawater mixed concrete is less durable than tap water mixed concrete.

3.2 Corrosion condition of reinforcements in concrete

(1) Corroded area

Fig. 5 shows the ratio of corroded area of reinforcements. Each value is the mean value of the measured values of three to five specimens of reinforcement under the same conditions. The following results were obtained.

· Rust generation was observed in the reinforcements in the concrete with portland cement at about 20 years of aging even with a covering of 7 cm.

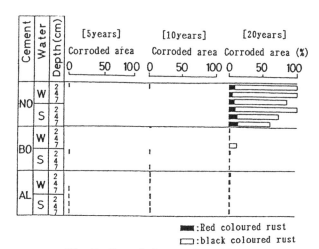

Fig. 5 Corroded area of steel bars

• Blast furnace slag cement and aluminous cement were more resistant to the corrosion of reinforcement than portland cement.

• Differences in the type of cement had a large effect on the corroded area.

• The use of seawater as the mixing water did not affect the corroded area very much.

• The effect of differences in covering thickness less than 7 cm on the corroded area was smaller than that of the differences in the type of cement.

(2) Judgment of the grade of passivity

The results of the grade of passivity of the reinforcements, as judged from the anodic polarization curve, are presented in Table 6.　From the table, it can be seen that as in the tendencies of the corroded area, the effect of the type of cement was more significant than that of the mixing water or covering depth.　That is, the grade of passivity of the reinforcements in the concrete which used the blast furnace slag cement or aluminous cement had better values than portland cement.　Clearly, this corresponds to the small value of corroded area in the concrete with blast furnace slag group cement or aluminous cement.

Table 6　Grade of passivity

Cement	Water	Depth(cm)	Grade of Passivity	
			15years	20years
N 0	W	2	2	2
		4	2	2
		7	2	2
	S	2	1	2
		4	2	2
		7	2	2
M 0	W	2	2	2
		4	2	2
		7	2	2
	S	2	2	2
		4	2	2
		7	2	2
B 0	W	2	4	4
		4	4	5
		7	4	5
	S	2	4	4
		4	4	5
		7	4	5
A L	W	2	-	5
		4	-	5
		7	-	5
	S	2	-	5
		4	-	5
		7	-	5

(3) Time–dependent change in mix potential with time

A typical change in mix potential with time is shown in Fig. 6. In short–term aging, there was a small difference between the values for seawater mixed and those for tap water mixed, but in long–term aging the values for seawater mixed were the same as those for tap water mixed. The tendencies of the changes in the potential differ depending on the type of cement used is considered to be mainly caused by differences in the oxygen amount around the reinforcement.

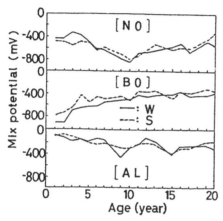

Fig. 6 Time–dependent change of mix potential

3.3 Chloride ions in the concrete

Fig. 7 shows the water soluble chloride ion as a weight percentage for mortar.

In N0, the chloride ions in the concrete are thought to nearly reach a condition of saturation at around 10 years of aging. On the other hand, the chloride content in

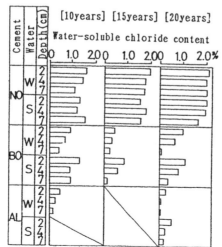

Fig. 7 Water–soluble chloride content in concrete

B0 and AL was much lower than that in N0, the chloride ion content at the age of 20 years was only 30% of that of N0 on the average.

As for the effect of the use of seawater as the mixing water on the chloride ion content in the concrete, no effect was recognized on N0; however, B0 and AL were affected a little because the amount of chloride penetration from the seawater is small. However, this effect was much smaller than that of cement type.

As for the effect of cover thickness on chloride ion quantity, there was no effect with N0. In contrast with B0 and AL, the effect of the cover thickness was remarkable; if a sufficient cover thickness is used a cover can prevent the penetration of salt.

Fig. 8 shows the relationship between the water soluble chloride content around reinforcements and the corroded area of reinforcements. As shown in Fig. 8, when the water soluble chloride content exceeds 1.5% in mortar weight, there was a clear tendency towards an increase in the corroded area.

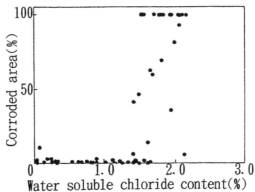

Fig. 8 Relationship between chloride content in concrete
and corroded area of steel bars

3.4 Diffusion coefficient of oxygen

Measured results of the diffusion coefficient of oxygen , which controls the cathodic reaction. Each value is an average value of samples taken from the top and the bottom of the test specimens.

As shown in the figure, mixing water had no effect on the diffusion coefficient of oxygen, but there were the recognizable effects of the type of cement. The average diffusion coefficients for each type of cement in the case saturation of 0% were as follows:

portland cement   : 9.35×10^{-4} to 10.57×10^{-4} cm^2/s
blast furnace slag cement   : 6.99×10^{-4} to 9.06×10^{-4} cm^2/s
aluminous cement   : 26.80×10^{-4} cm^2/s

The average diffusion coefficient for blast furnace slag cement was a little lower than that for portland cement. In contrast, that for aluminous cement was remarkably higher. This can be explained by high porosity due to the conversion of mortar of aluminous cement.

4 Conclusion

The results of the exposure test for a 20 year period under a marine environment (tidal zone) can be summarized as follows.

1. The compressive strength of concrete increases up to about 5 years of aging, after this gradually decreases. There were no differences in compressive strength due to differences in mixing water with long– term aging.

2. Based on the corroded area and the grade of passivity, it was clear that the corrosion of the reinforcements in the concrete using portland cement progresses up to 20 years of aging. The corrosion in concretes with blast furnace slag cement and aluminous cement was less than that in concrete with portland cement. There was almost no recognizable effect on the corrosion of reinforcements due to differences in mixing water.

3. As for preventing the penetration of salt, the concretes using blast furnace slag cement and aluminous cement were much better than these using portland cement.

4. There was no effect of mixing water on the diffusion coefficient of oxygen in the concrete. On the other hand, there was a recognizable effect of the type of cement used on the diffusion coefficient; the concrete with the lowest diffusion coefficient was blast furnace slag cement.

5. Within the scope of this test, even when gypsum phosphate with 2% in SO_3 weight was added to ordinary portland cement and blast furnace slag cement B– class, there was no recognizable effect on the durability of the concrete.

6. In this series of experiments for 20 years, concrete using blast furnace slag cement and aluminous cement show very excellent performance against sea– water attack and corrosion of reinforcement. Furthermore, it was found that sea water as mixing water of concrete has little effect on the durability of concrete and corrosion of reinforcement, when concrete has been exposed in tidal zone for long time.

5 References

1. The Cement Association of Japan (1969) Cement with different quantities of SO_3 for "Research into the seawater– proofing of concrete," *Cement– Concrete*, No.273, pp. 26– 27. (in Japanese)

2. Nobuaki Otsuki (1985) Research on the influence of chloride on corrosion of embedded steel bars in concrete, *Report of Port and Harbour Research*

Institute, No. 3, Vol. 24. (in Japanese)

3. Japan Concrete Institute (1987) Test methods and standards for the corrosion and corrosion- proofing of concrete structures (draft), pp. 23–58. (in Japanese)

4. Japan Concrete Institute (1989) Report by the Durability Examination Research Committee, pp. 29–36. (in Japanese)

5. Gjorv, O. E. (1971) Long–time durability of concrete in seawater, *ACI Journal*, Vol. 68, pp. 60–67.

6. Shigejiro Mori et al (1974) Test results of 20 years of aging on concrete using many types of cement, *Onoda–Kenkyu Hokoku*, No. 92, Second book, Vol. 26, pp. 106–116. (in Japanese)

7. Hisao Ohyama (1989) Seawater-proofing of concrete using many types of cement, *MMCC research report*, No. 2, pp. 21–36. (in Japanese)

15 SHOTCRETE FOR REPAIRS: ISSUES AND CHALLENGES

N. BANTHIA, R. HEERE and H. ARMELIN
University of British Columbia, Vancouver, Canada

Abstract
Repair and rehabilitation of the deteriorating infrastructure is one of the major challenges facing the industrialized nations of the world today, and there is an urgent need to develop ways of producing cost effective repairs which are also durable and aesthetically pleasing. Of the various techniques available, shotcreting is one of the most promising. However, in spite of its long history of use, many aspects of shotcrete remain poorly understood and major research initiatives are necessary. This paper discusses some of these issues and describes the authors' current attempts at understanding shotcrete through laboratory and field studies.
Keywords: Dams, fiber reinforcement, repairs, rebound, shotcrete

1 Introduction

In most developed countries, the infrastructure is rapidly deteriorating or needs urgent upgrading. Demolition of these structures to construct new ones, unfortunately, is not an option for Governments exercising severe fiscal restraint; repair and rehabilitation is the only solution. Unfortunately, our current repair and rehabilitation techniques are largely inadequate and much research is needed to develop rapid, cost effective and durable ways of repairing these deteriorating structures.

Shotcreting is now considered as one of the most promising repair techniques. Defined normally as, "mortar or concrete pneumatically projected at high velocity on to a surface", shotcrete is extensively used in tunnel linings, slope stabilization works, culverts, sewers, dams, canals, swimming pools and in the repair and rehabilitation of highway, railway and marine structures. Because of the very different mix proportions and its unique compaction technique, shotcrete is fundamentally different from concrete and as such our accumulated understanding of traditional concrete can not be

Integrated Design and Environmental Issues in Concrete Technology. Edited by K. Sakai. Published in 1996 by E & FN Spon, 2–6 Boundary Row, London, SE1 8HN, UK. ISBN 0 419 22180 8.

directly applied to shotcrete. It is understandable, therefore, that many aspects of shotcrete to this day remain insufficiently understood.

The purpose of this paper is to discuss some of the issues and challenges that relate to the production, placement and performance of shotcrete when used as a repair material.

2   Dry and wet-mix shotcrete

Shotcrete can be produced via two processes - dry-mix and wet-mix. In the dry-mix process, as the name indicates, dry pre-mixed materials are blown through a delivery hose to a nozzle where the water is added. In the wet-mix process, on the other hand, mix ingredients including water are premixed and then conveyed through a hose to the nozzle where the compressed air is added. Through the nozzle, the material is shot at a high velocity on to the receiving surface where it compacts under its own velocity and gains strength.

As will be discussed in a greater depth later, part of the material shot on the application surface rebounds and the *rebound* consists mostly of large aggregate particles. Consequently, in dry-mix shotcrete, where the rebound problem is more noticeable, aggregate particles larger than 4.75 mm (ASTM No. 4 sieve) are rarely used. In wet-mix shotcrete, on the other hand, where rebound is not such a big concern, pea gravel or larger aggregate particles (up to 19 mm) are commonly used. Water-cementitious ratios in dry and wet mix shotcretes are also significantly different. In the dry-mix, since one can control the water at the nozzle, only sufficient water which will allow (partial) hydration of cement is added. In the wet-mix process, on the other hand, the plastic mix has to be pumped and conveyed towards the nozzle and hence must be workable enough. In spite of the use of highly superplasticized mixes and high quality shooting equipment, wet-mix shotcrete almost always ends up with a water-cement ratio higher than the dry-mix shotcrete, which makes the wet mix shotcrete significantly more prone to shrinkage related cracking, more porous internally and also significantly more permeable. An excellent description of the fundamentals of shotcrete is given by Warner (1).

3   Plain and reinforced shotcrete

Plain shotcrete is as brittle as concrete and equally weak in tension. Reinforcement in one form or the other is, therefore, necessary. Conventional reinforcing bars interfere significantly with the placement of shotcrete and this alone has been the reason for number of unsatisfactory placements in the past. Galvanized welded wire mesh (openings of about 50 to 100 mm square) is often provided with or without conventional reinforcing bars.

The difficulties experienced in shotcreting in the presence of reinforcing bars has been the major reason behind the success of fiber reinforced shotcrete. Short fibers (15-30 mm long) of steel or other polymeric materials are added to the mix along with the other ingredients and the mixes are shot on to the application surface without any other conventional reinforcement. Fiber reinforced shotcrete has its distinct advantage in that it follows the contour of an uneven substrate—something not possible with welded wiremesh—and when properly placed and cured, it is tough, durable and dimensionally stable.

4 The rebound problem

The rebound of the materials during the process of shooting is a real concern in shotcrete. Rebound, as can be expected, is a function of the process adopted and it can be much higher in the dry-process (usually 30 to 50%) as compared to that in the wet process (usually 7 to 15%). Rebound is a also a function of the size and specific gravity of the particles shot and the stiffness of the surface on which the material impinges. Because the rebound is largely composed of coarse aggregate particles, it causes the in-place shotcrete to be composed of a higher cement content than the design mix. For example, a dry-mix shotcrete designed with a cement content of about 400 kg/m^3 results in an "in-situ" cement content of the order of about 600 to 700 kg/m^3. Besides the obvious loss of material and productivity, aggregate rebound results in a final product on wall which, due to its high cement content, is prone to large dimensional changes especially at the early ages. In the case of steel-fiber reinforced dry-mix shotcrete, the rebound of fibers tends to be even higher than the aggregate particles, and depending upon the fiber geometry, more than 60% of the fibers may be lost in the process (see Table 1). Clearly, this results in an in-situ fiber content which is much lower than intended and a hardened shotcrete with significantly diminished post-crack toughness.

Table 1. Fiber rebound data in shotcrete (Ref. 2-3)

Fiber	Fiber Description	Fiber Rebound by Volume* (%)	
		Dry-Mix	Wet-Mix
F1[✧]	Hooked-ends, circular section, 0.5 mm diameter, 30 mm long.	49.1	18.3
F2	Hooked-ends, rectangular section, (0.45 × 0.53 m), 32 mm long.	34.8	11.5
F3	Crimped fiber, crescent section (0.50 × 1.35 mm), 32.5 mm long.	69.0	12.5
F4	Milled fiber, irregular section, 32.5 mm long.	48.2	17.0
F5	Twisted along length, rectangular section (0.25 × 1.12 mm), 25.5 mm long.	77.8	11.8

[✧] F1...F5 are steel fibers available commercially with different geometries (see Ref. 2,3 for further details).

* Calculated as $[(V_f)_o-(V_f)_w]/(V_f)_o$ where $(V_f)_o$ = original fiber volume fraction, and $(V_f)_w$ = actual fiber volume fraction on wall.

Although rebound has long been recognized as a major problem in shotcrete, a generalized theory of rebound capable of explaining the various observations and modeling the various inherent processes has never been developed. One of the main reasons for this lack of understanding is the high velocities involved in shotcreting which makes it almost impossible to accurately model the process. So far, lack of appropriate tools has deterred an assessment of velocity and acceleration of aggregate and fiber particles in the shotcrete stream and has not allowed the study of particle entrapment, embedment or rebound. In addition, from a theoretical point of view, the straight-forward assumption of fresh concrete being a Bingham fluid is invalid in studies of rebound since it would assign no reaction to the shotcrete substrate and hence predict zero rebound.

Attempts are currently underway at the University of British Columbia (4) to study the rebound mechanisms in steel fiber reinforced shotcrete using high-speed camera running at a speed of 1000 frames per second. Fundamental studies of particle motion include studying the motion of aggregate and fiber particles in an air stream and to characterize their velocities and accelerations as a function of particle size, fiber geometry and the air pressure. Results so far indicate that particle velocities are characterized by a great scatter and are related to particle size such that smaller particles acquire greater cruising velocities than the larger particles (Figure 1). Figure 2 shows photographs of fibers captured in the stream. From studies of this kind, one can not only learn of the kinematics of fiber motion in the shotcrete stream, but also relate the various forces acting on the fiber to its translatory and rotatory motions along the three axes. This information serves as a precursor to the eventual behavior of the particle when it strikes the application surface and either embeds or rebounds. Both these stages of particle behavior, before and after the contact with the substrate, are clearly dependent on the mix design, presence of mineral and chemical admixtures, magnitude of air pressure, fiber geometry, aggregate angularity, etc., and must somehow be incorporated in the theoretical model. Ultimately it is hoped that this fundamental understanding of the rebound mechanisms will lead to mixes and shooting techniques that produce minimal fiber and aggregate rebound.

5   Shrinkage cracking in shotcrete and the influence of fibers

The fact that rebound in shotcrete consists mostly of large aggregate particles implies that the aggregate content in the in-place shotcrete is very low and accordingly the cement content is very high. This coupled with the fact that the high surface area of shotcrete permits an increased loss of moisture under low humidity and high temperature environments indicates that shotcrete will be significantly more susceptible to shrinkage related cracking. It is a well established fact that fibers in shotcrete not only delay the formation of plastic shrinkage cracks initially, but also develop a sufficiently strong bond with the plastic matrix early in the process and restrain the shrinkage cracks from widening. This is illustrated in Figures 3a and b where the effectiveness of three steel macro-fibers (with different geometries) and four micro-fibers in restricting the crack width in concrete is illustrated (5). There is a need to study shotcrete with reference to the effectiveness of fibers in restricting crack widths

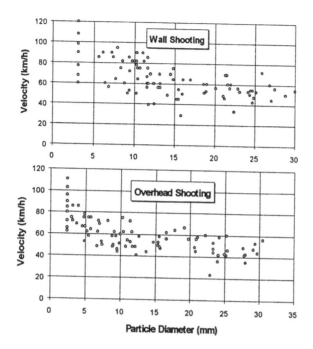

Fig. 1. Aggregate velocity in the shotcrete stream for wall and overhead shooting. Notice that smaller aggregate particles acquire lower cruising velocities.

as a function of environmental conditions of temperature and humidity. Some such studies are currently underway (6).

6 Toughness characterization in fiber reinforced shotcrete

As mentioned previously, plain shotcrete is as brittle as traditional concrete and fiber reinforcement provides an enhanced toughness or deformability. Depending on the fiber volume fraction, the fracture toughness (measured as critical stress intensity factor, K_{IC}) of fiber reinforced shotcrete can almost be an order of magnitude higher than plain shotcrete. This imparts fiber reinforced shotcrete a significantly improved energy absorption capability and an enhanced resistance to impact and fatigue loads.

One major issue facing the industry at the moment is how to characterize the toughness of fiber reinforced shotcrete or concrete for that matter. The currently available methods (ASTM C1018 or JSCE SF-4) are both unacceptable (2,3,7,8) and provide misleading toughness parameters that are either operator dependent and/or irrelevant to field performance with only limited usefulness in design. Concerted efforts are needed on parts of both the industry and the academia to devise rational methods of characterizing the toughness of fiber reinforced shotcrete which will not only be objective and free from spurious influences but also produce parameters which the engineers can design with.

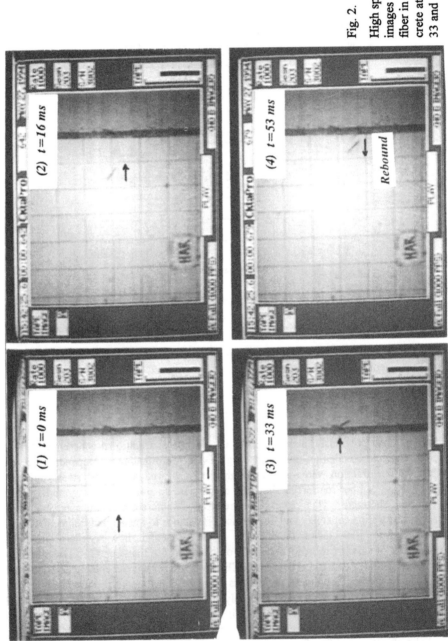

Fig. 2.

High speed images of a fiber in shotcrete at 0, 16, 33 and 53 ms.

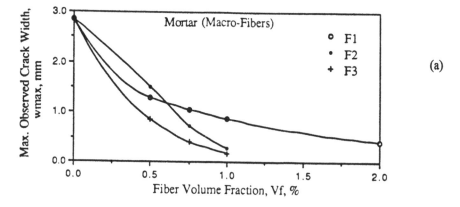

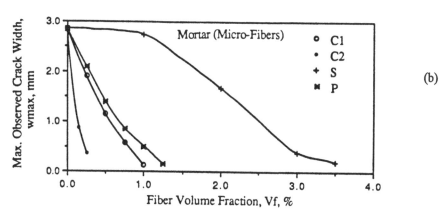

Fig. 3. Restrained shrinkage cracking as controlled by: (a) Macro-fibers; and (b) Micro-fibers (*F1..F3 are shotcrete steel fibers with different geometries; C1 and C2 are pitch-based carbon fibers with different lengths, P is polypropylene fiber and S is steel micro-fiber*).

7 Performance and durability of shotcrete in BC Hydro dams

During 93-94, an extensive investigation (9) was undertaken to evaluate shotcrete repairs in four major dams owned by BC Hydro which is the main producer of electricity in Canada's pacific province of British Columbia. About 90 percent of the utility's generating capacity (9.3 out of 10.5 GW) is based on hydro power emanating from 43 dams most of which are concrete gravity or concrete buttress structures. Nearly one fourth of these dams are older than 50 years with concrete having no air-entrainment and water/cement ratios as high as 0.6 to 0.7 with poorly graded aggregates. Leakage and freeze-thaw deterioration is common in these dams which necessitated resurfacing and the earliest shotcrete repairs were carried out in 1953. In the investigation reported here, shotcrete was surveyed for performance and durability in four dams: the Buntzen dam, Ruskin dam, Stave dam and the Jordan dam all within

100 km of Vancouver. Extensive field and laboratory studies were performed at various sites. Table 2 presents the characteristics of some of the durable repairs

Table 2. Shotcrete repairs that performed satisfactorily

	Buntzen Dam	Ruskin Dam	Jordan Dam, Slab, Upstream	Jordan Dam, Buttresses
Substrate Concrete				
Poured [year]	1911 to 1914	1929 to 1930	1912 to 1913	1912 to 1913
Compressive Strength [MPa]	25	14 (at 28 days)	26	18
Max. Aggregate Size [mm]	~40	150	30	30
Cement : Aggregate Ratio by Mass	1 : 8	1 : 9.8	1 : 6	1 : 9
Water : Cement Ratio by Mass	?	?	~0.65	~0.65
Ultrasonic Pulse Velocity [km/s]	~4	~4.5	4.6	?
Remarks	contaminated with soil and wood		large air voids	large air voids
Environment				
Annual Precipitation [mm]	2900	3000	3500	3500
Orientation of Repair	West	West	North	South
Approximate Temperature Range [°C]	-15 to +30	-20 to +30	-20 to +25	-20 to +30
Annual Freeze-Thaw cycles	46	38	40	40
Surface Preparation	generally poor			
mechanical	?	trolley-mounted rotating cutheads and scrubbers	chipping	chipping
sand/water blasting	?	dry sandblasting, flushing with water prior to shooting	sandblasting or water blasting	water blasting
Shotcrete Mix				
Year of Repair	1965	1973	1969	1990
Type	dry-mix	dry-mix	dry-mix	wet-mix
Cement : Aggregate Ratio by Mass	?	1 : 3.5	?	1 : 4.2
Max. Aggregate Size [mm]	5	5	5	5
Steel Fiber Content [kg/m^3]	0	0	0	0
Silica Fume Content [kg/m^3]	0	0	0	50
Curing	?	soaker hoses	soaker hoses	7 days moist
Hardened Shotcrete				
Thickness [mm]	80 to 130	100 to 250	75 to 150	> 65
Reinforcement	4 mm steel wire mesh	5 mm galvanized steel wire mesh 100 mm x 100 mm	4 mm steel wire mesh	steel wire mesh 150 mm x 150 mm
Compressive Strength [MPa]	50	> 60	33	51
Paste Content - ASTM C457 [%]	32	41	38	35
Air Content - ASTM C457 [%]	5.6	4.8	1.9	4.0
Air Void Spacing - ASTM C457 [mm^{-1}]	0.93	0.79	1.0	0.17
Permeable Voids - ASTM C642 [%]	17 to 20	11 to 15	16	13
Boiled Absorption - ASTM C642 [%]	7.8 to 9.0	4.9 to 6.8	6.8	5.6
Ultrasonic Pulse Velocity [km/s]	4.3	4.7	4.6	4.7
Tensile Bond Strength [MPa]	1.4	~1 to 2.4	1.3	1.1

encountered in this survey and Table 3 describes some of the less durable ones. Figure 4 shows a typical cracking pattern resulting from shrinkage at the Jordan dam and Figure 5 shows freeze-thaw damage in shotcrete on the horizontal face of walkway at the Buntzen dam. Shotcrete in this case was of poor quality due to downward shooting and was particularly more susceptible to frost damage due to prolonged saturation. Figure 6 shows the shotcrete repairs on the spillway at the Ruskin dam showing vertical cracks developed along construction joints that are approximately 20 m apart. Severe delamination could also be noticed in steel fiber reinforced shotcrete at Stave dam as seen in Figure 7 where water could migrate through a rain trap on top of the retaining wall to the substrate-shotcrete interface and induce severe cracking upon freezing.

Table 3. Unsatisfactory shotcrete repairs on B.C. dams

	Stave Dam	Ruskin Dam, Retaining Wall	Jordan Dam Downstream /S Face
Substrate Concrete			
Poured [year]	1922 to 1923	1929 to 1930	1912 to 1913
Compressive Strength [MPa]	28	14 (at 28 days)	26
Max. Aggregate Size [mm]	40	150	30
Cement : Aggregate Ratio by Mass	1 : 7.5	1 : 9.8	1 : 6
Water : Cement Ratio by Mass	0.7 to 0.9	?	~0.65
Ultrasonic Pulse Velocity [km/s]	4.4	~4.5	4.6
Remarks	fraction of mid-sized aggregates seemed small		large air voids
Environment			
Annual Precipitation [mm]	3000	3000	3500
Orientation of Repair	North-West	West	North
Temperature Range [°C]	-20 to +30	-20 to +30	-20 to +25
Annual Freeze-Thaw cycles	38	38	40
Surface Preparation			
mechanical	chipping	trolley-mounted rotating cutheads and scrubbers	chipping
sand/water blasting	water blasting	dry sandblasting, flushing with water prior to shooting	sandblasting or water blasting
Shotcrete Mix			
Year of Repair	1985	1973	1970
Type	dry-mix	dry-mix	dry-mix
Cement : Aggregate Ratio by Mass	?	1 : 3.5	?
Max. Aggregate Size [mm]	10	5	5
Steel Fiber Content	60	0	0
Silica Fume Content	(7% by cement mass)	0	0
Curing	none	soaker hoses	?
Hardened Shotcrete			
Thickness [mm]	<60	75 to 150	~20
Reinforcement	none	5 mm galvanized steel wire mesh 100 mm x 100 mm	apparently none
Compressive Strength [MPa]	~40 (by Schmidt Hammer)	> 60	21 to 64
Paste Content - ASTM C457 [%]	32	41	39
Air Content - ASTM C457 [%]	6.2	4.8	2.6
Air Void Spacing - ASTM C457 [mm^{-1}]	0.45	0.79	0.26
Permeable Voids - ASTM C642 [%]	18	11 to 15	?
Boiled Absorption - ASTM C642 [%]	8.1	4.9 to 6.8	?
Ultrasonic Pulse Velocity [km/s]	4.2	4.7	4.6
Tensile Bond Strength [MPa]	0.2	0 (shotcrete debonded)	1.3
Problems /Failure	shotcrete delaminating and cracking, substrate concrete decaying	shotcrete completely delaminated, freeze-thaw attack on horizontal areas, substrate concrete decaying	shotcrete delaminated, severe shrinkage cracking

Fig. 4. Typical shrinkage cracking in shotcrete at the Jordan dam.

Fig. 5. Freeze-thaw damage in shotcrete at the Buntzen dam.

Fig. 6. Shotcrete repairs on the spillway at the Ruskin dam showing vertical cracks along construction joints.

Fig. 7. Delamination in steel fiber reinforced shotcrete at the Stave dam.

Based on this study, the following recommendations for producing durable shotcrete repairs, particularly as applied to hydraulic structures, may be made:

7.1 Surface preparation

A surface receiving shotcrete repair must be dimensionally stable with respect to its environment to avoid overstressing of the repair shotcrete. A careful substrate surface preparation is essential for good bond between the substrate concrete and shotcrete. e substrate surface should be roughened (by chipping or hydrodemolition) to allow for a good interlocking with the shotcrete and be free from all loose material. Immediately before applying shotcrete, the substrate surface should be in a dust-free and "saturated-surface-dry" condition. Finally, the moisture content and porosity of the substrate should not significantly change the w/c-ratio of the freshly applied shotcrete. Any joints in the substrate will require expansion joints in the shotcrete. One must assure that the substrate will not remain critically saturated in service, otherwise only a few cycles of freezing will delaminate the repair.

7.2 Shotcrete material

The mix to be shot should be uniformly mixed and stable. In wet-mix shotcrete, the mix should have appropriate consistency for easy pumping and should not segregate or bleed while being pumped or cause abrasion in the equipment. It should also not set while being inside the equipment. Properly proportioned and shot material will not leave shadows behind reasonably sized reinforcement. Upon hardening, shotcrete should develop properties such as strength, moduli, coefficient of thermal expansion, etc., which are not very different from those of the substrate. The shotcrete should bond permanently to the substrate, have high abrasion resistance and be freeze-thaw durable. Mixes containing about 400 to 450 kg/m^3 of cement, 50 kg/m^3 of silica fume and about 80 kg/m^3 of deformed steel fibers appear to perform satisfactorily in most cases. In the wet-mix shotcrete, 0.1 liter of air-entraining agent and about 2 liters of high-range water reducer per 100 kg cement produces the most satisfactory results. Latex or accelerators should generally be avoided in shotcrete for repairs on hydraulic structures.

7.3 Application technique

Rebound and dust development should be avoided to the extent possible. The technique adopted should allow shooting from all positions and variable nozzle distances. The distance of the nozzle from the substrate, its angle and its motion are dependent on the climatic conditions (wind, etc.) at the time of shooting and only an experienced nozzleman can adjust to changes in site conditions to produce durable shotcrete. If moisture and frost are a concern, the repair shotcrete should be at least 100 mm thick. Experience indicates that shotcrete jackets 250 mm in thickness are sufficiently self-supporting even in the event of significant substrate deterioration.

8 Closure

Shotcreting is a repair technique with a great deal of promise. In spite of its long history of use, many aspects of shotcrete remain poorly understood and concerns with its performance in the field persist. Major research initiatives are needed to understand the fundamental kinematic processes involved in the placement of shotcrete, develop better tools for its characterization and devise techniques aimed at enhancing its performance in the field.

9 Acknowledgments

The continuing financial support of the Natural Sciences and Engineering Research Council of Canada is gratefully acknowledged. Thanks are also due to Dr. D.R. Morgan for his constant encouragement and valuable advice.

10 References

1. Warner, J. (1995) Understanding Shotcrete-The Fundamentals, ACI Concrete International, pp. 59-64
2. Banthia, N. et al. (1992) Influence of Fiber Geometry in Steel Fiber Reinforced Dry-Mix Shotcrete, Concrete Int.: Design and Construction, ACI, Vol. 14, No. 5, pp. 24-28.
3. Banthia, N. et al. (1994) Influence of Fiber Geometry in Steel Fiber Reinforced Wet-Mix Shotcrete, Concrete Int.-Design and Construction, ACI, Vol. 16, No. 6, pp. 27-32.
4. Armelin, H. (1995) Kinematic Studies in Fiber Reinforced Dry-Mix Shotcrete, Ph.D. Thesis, *in preparation*, The University of British Columbia, Vancouver.
5. Banthia, N. and Azzabi, M. and Pigeon, M. (1993) Restrained Shrinkage Cracking in Fiber Reinforced Cementitious Composites, Materials and Structures RILEM (Paris), 26 (161), pp. 405-413.
6. Campbell, K. (1995) Restrained Shrinkage Cracking in Fiber Reinforced Shotcrete, M.A.Sc. Thesis, *in preparation*, The University of British Columbia, Vancouver.
7. Morgan, D.R., Mindess, S. and Chen, L. (1995) Testing and Specifying Toughness For Fiber Reinforced Concrete and Shotcrete, Proceedings of Second University-Industry Workshop on Fiber Reinforced Concrete and Other Advanced Composites (Eds., N. Banthia and S. Mindess), Toronto, pp. 29-50.
8. Banthia, N. et al. (1995) Test Methods of Flexural Toughness Characterization: Some Concerns and a Proposition, American Concrete Institute, Materials Journal, 92(1), pp. 48-57.
9. Heere, R. (1995) A Survey of Shotcrete Repairs in BC Hydro Dams, M.A.Sc. Thesis, The University of British Columbia, Vancouver.

16 LIFE EXTENSION OF CONCRETE STRUCTURES USING ADVANCED COMPOSITE MATERIALS

S.A. SHEIKH
University of Toronto, Toronto, Canada

Abstract

Repair, rehabilitation, retrofitting or life extension of existing structures has gained new significance in the light of current environmental awareness of our society. The importance of this engineering task can be estimated from the fact that in North America alone the rehabilitation costs associated with the highway infrastructure alone is estimated at $130 billions (US). Repair of other structures such as airports, buildings, etc. would require similar resources.

Various techniques for rehabilitation are used in practice which employ traditional materials such as steel and concrete. Use of advanced composite materials (ACM), also known as fibre reinforced plastics (FRP), for life extension of structures is described here as an alternative to the traditional techniques. Under certain circumstances use of ACM can provide a more economical and technically superior solution to a problem compared to most other existing methods.

In this paper some examples of structural damage are given with the emphasis that most of this damage could be avoided or minimized with good engineering practice. After a general introduction, a brief description of ACM properties is provided. Some results from the available experimental research on the repair of beams and columns using ACM are given to familiarize the reader with the state-of-the-art. Behavior of beams and columns strengthened by ACM is discussed with examples and a research study on the corrosion and repair of columns is briefly detailed.

1 Introduction

Chemicals such as sulfates, acids and alkalis, variations in temperature between freezing and thawing and changes in moisture contents affect the durability of paste

Integrated Design and Environmental Issues in Concrete Technology. Edited by K. Sakai. Published in 1996 by E & FN Spon, 2–6 Boundary Row, London, SE1 8HN, UK. ISBN 0 419 22180 8.

and aggregate in concrete. Corrosion of steel in concrete structures as a result of chemical attack is one of the main causes of concrete degradation. Other than the harsh environment, unsound construction practices, design inadequacies, a lack of quality control combined with minimal maintenance have resulted in a vast inventory of structures that are either unsafe to use at all or are in a dire need of repair for continued usage.

In North America alone there are over half a million highway bridges of which more that 200,000 are deemed deficient. More than 5,000 bridges are closed because they are unsafe. Every year about 200 spans fail partially or completely, sometime with tragic loss of life [1]. The rehabilitation cost of these structures is estimated at US $130 billions [1, 2]. Fig. 1 shows a damaged bridge in Canada. A significant observation is that the damage to the structure is mostly in the areas directly below the expansion joint in the bridge deck. The expansion joint over a period of time deteriorates to an extent that it provides a passage for the de-icing salt solution to pass through and flow over the supporting girders and columns.

Figs. 2 and 3 show areas of an underground garage of a three year old high-rise residential building in which extensive cracking has occurred. The cracks started appearing only a few weeks after casting of concrete. Fig. 2 shows crack pattern in the West Wall and Fig. 3 displays the underside of a part of the slab. Whereas most cracks can be classified as hairline, several cracks are between 0.4 mm and 0.75 mm in width. In perhaps the most damaged zone of the wall around grid lines 8 and 9 the average with of cracks is about 0.35 mm and the approximate strain is 0.2×10^{-3}. At the North end of the wall, horizontal cracks at the location of the maximum moment run for about 30 m.

Fig. 1. Damaged bridge along highway 401 in Ontario

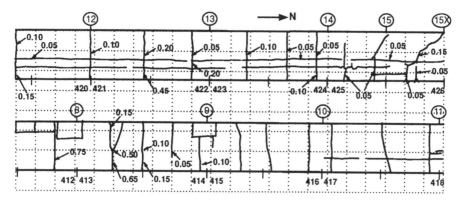

Fig. 2. Cracks on West wall (partial)

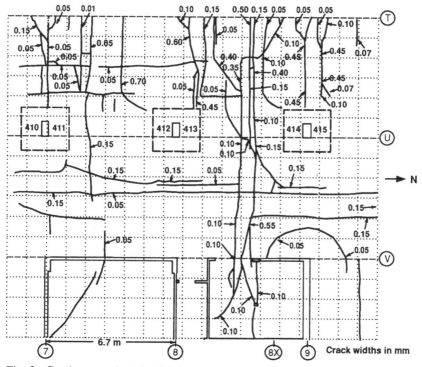

Fig. 3. Cracks on underside of slab (partial)

The slab area between grid lines 8 and 9 shows a dense crack population under service load (Fig. 3). In the east-west direction the average crack width is 0.24 mm, average crack spacing is approximately 600 mm and the average strain can be approximated to be 0.4×10^{-3}. The north-south crack between grid lines U and V which covers almost the entire 80 m length of the structure is about 0.5 mm thick at several locations. Whereas shrinkage and temperature variations have played a large role in the development of most of the cracks, some of the cracks can be attributed directly to the moment generated by the load.

The North American codes [3, 4] do not specify crack width limitations for service load conditions. However, the European code [5] recommends that for moderate exposure conditions, the average crack width should not exceed 0.15 mm and no more than 5% of the cracks should exceed a width of about 0.25 mm. These limits are exceeded by a wide margin at several locations in the structure. The structural design generally meets the code requirements. Field tests indicated that the constructed structure is built reasonably according to the drawings and specifications. The only notable exception was the excessive slump of concrete and lower than the specified strength on several occasions. Slump as high as 140 mm instead of the specified 75 mm and strength as low as 75% of the specified strength were observed most likely due to the water added at the site which resulted in excessive shrinkage. Fig. 4 shows a qualitative relationship between strength and durability parameters [6]. Although strength of Concrete A is only slightly higher than that of Concrete B, its durability is significantly better indicating that the quality of concrete is not accurately reflected by the strength measurements alone.

In the examples given above and many other instances factors causing the structural distress include design flaws, construction practices, lack of quality control and site supervision. In many cases these adverse effects can be avoided entirely and in others their effects can be minimized with minimal additional cost. Site supervision has become an ignored part of engineering particularly on small projects with the result that despite advances in new materials and techniques a large number of structures still experience distress at an early age which provides an opportunity for further accelerated deterioration such as steel corrosion and concrete spalling and delamination.

Conventional repair techniques which involve traditional materials such as steel and concrete have been successfully employed for decades. However, in situations where the conventional techniques are found cumbersome and require closing of the entire damaged area of the structure, alternative repair techniques are needed which can allow regular activities to continue with minimal disruption. Results from recent work on advanced composite materials (ACM) and its applications in repair and rehabilitation are very promising in this regard. In addition to being less labor intensive, these composites are lighter, stronger and corrosion resistant.

2 Properties of advanced composite materials

When two or more materials are combined on a macroscopic scale to form a new material, the resulting product is a composite material. Typical fibrous composites are obtained using glass, carbon or aramid fibres in a polymer matrix. The main function of the matrix, the binding material of the composite, is to support and protect the load carrying fibres. In addition it also assumes a role of transmitting the applied loads to fibres basically through the shearing stresses that develop at the interface between fibre and matrix.

The most common matrices are resinous materials such as polymers which include various types of epoxies, polyesters, and epoxides [7, 8]. Tensile strength of these resins vary between 40 MPa and 100 MPa and the modulus of elasticity is in the range of 2500 to 5500 MPa. Some low strength (20-40 MPa) matrices have stress-strain characteristics which display inelastic deformation before rupture. Higher

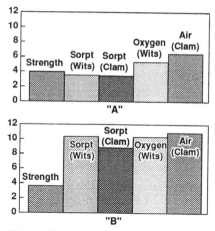

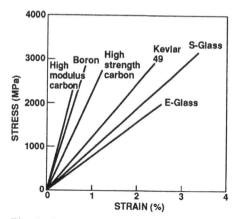

Fig. 4. Concrete quality parameters

Fig. 5. Stress-strain curves for typical fibers

strength (50-100 MPa) matrices generally fail in tension with little deformation beyond the elastic stage. Fig. 5 shows the stress-strain behavior of various fibres under tension. The brittle failure lacking any substantial inelastic deformation is typical of most of these fibres.

Table 1 illustrates a summary of the mechanical properties of some typical fibre-reinforced plastics and compares these with those of steel. The stiffness values for most FRPs are considerably less than that of steel. In addition to the proportions of the constituents and their mechanical properties, other factors such as fibre orientation

Table 1. Comparative mechanical properties of ACM and steel

Material	Fibre content [# by weight]	Density [kg/m³]	E_t [MPa]	f_t [MPa]
Unidirectional GFRP/Polyester Laminate	50-80	1600-2000	20 000-60 000	400-1250
GFRP/Polyester Randomly Oriented Hand Lay-up	25-25	1400-1600	6000-11 000	60-180
GFRP/Polyester Matched Metal Dye	25-50	1400-1600	6000-12 000	60-200
GFRP/Polyester Woven Rovings Hand Lay-ups	45-62	1500-1800	1200- 2400	300-350
Sheet Moulding Compound, Unidirectional Laminate	20-25	1750-1900	9000-13 000	60-100
Aramid-Epoxy	50-80	-	70 000-80 000	1000-1400
Carbon/Epoxy	70-80	1900-2000	120 000-180 000	800-1200
Steel		7800	200 000	250-1000

within the matrix, fabrication process, and strength of the fibre-matrix bond also influence the properties of the composite materials.

3 Rehabilitation of structures with ACM

Deterioration of older structures due to loss of material properties, inferior design, action of climate, substandard construction as a result of revision of design codes, or increase in traffic loads in the case of bridges, often requires the invention of new rehabilitation techniques which are more feasible than the existing ones. Strengthening of a structural element can sometimes be achieved with the installation of post tensioning cables. ACM prestressing cables can be installed as external reinforcement for beams, or as internal reinforcement in a box girder [9]. The technique of increasing the load-bearing capacity of bridges by bonding steel plates on the bottom face of girders can be adapted by using ACM instead of steel.

A rehabilitation or retrofitting technique for columns, originally proposed in U.S.A., consists of wrapping ACM sheets around the columns. This technique should improve, among other things, structural behavior of columns under seismic loading. This approach is very similar to the reinforcement of tall chimneys with a combination of ACM laminae and wires, and which has already been used extensively in Japan [8, 9].

3.1 Beams
In the past, concrete bridges were strengthened by additional beams, props, or external post-tensioning. In recent years, with the development of strong structural adhesives, plate-bonding techniques have been recognized to be an effective method of improving the performance of concrete structures under service loads. However, the weight of the steel plates makes them difficult to handle and restricts the size of plate that can be used. There are also problems such as corrosion of the steel-adhesive interface and plate separation due to high local interface bond stresses and peeling forces at the ends of the plates. FRP strengthening sheets could be a practical solution to these problem. It has been reported [10] that for a specific application, it was possible to replace 94 kg of steel by 4.5 kg of CFRP. Strengthening is achieved by bonding unidirectionally 0.3 mm to 4 mm thick CFRP or GFRP sheets to the beam using epoxy resin adhesives. Special attention must be paid to the development of shear cracks in concrete, which can lead to a premature peeling-off of the strengthening plates. Use of CFRP laminate also reduces the total crack width and evenly distributes the cracks.

Fig. 6 shows a comparison between repaired and control beams using two different repair schemes [11]. The figure also includes trace of preloading condition of the original beams. The beams were first damaged by loading to 85% of their flexural capacities. The beam P3 was repaired using 3 mm thick GFRP sheet bonded to the beam soffit. Beam P3J was repaired using the same 3 mm thick GFRP sheet manufactured in the form of a one piece I-jacket plate, which was glued not only to the beam soffit but also to the sides in the shear span. Fig. 6 shows that simple use of ACM sheet as tension reinforcement can recapture or even enhance the original beam capacity but reduces ductility resulting in brittle failure as a result of separation of ACM sheet from the beam. Connecting the ACM sheet to the beam with the help of

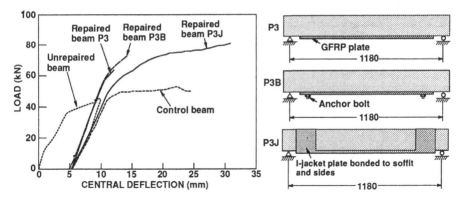

Fig. 6. Load-deflection behaviour of the beams tested by Sharif et al. [12]

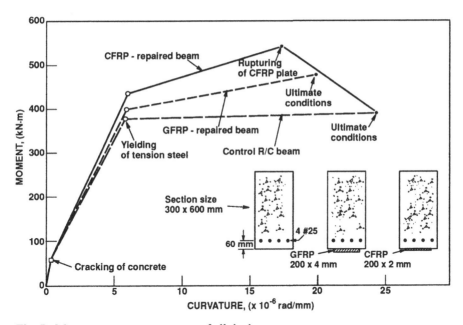

Fig. 7. Moment-curvature response of all the beams

especially designed bolts resulted in a slight improvement in the capacity but the failure was caused by diagonal tension in the shear zone. The I-jacket GFRP plate (Beam P3J) provided adequate anchorage system to eliminate plate separation as well as diagonal tension failure to develop ductile flexural behavior of the repaired beam.

3.2 Beam repair example

Fig. 7 compares the moment-curvature behavior of a regular RC beam section with those of ACM-retrofitted beams. Following material properties were used in the analysis: concrete compressive strength f'_c = 30 MPa, tensile strength = 3 MPa, steel yield strength f_y = 400 MPa and strain hardening is not considered. CFRP tensile strength = 800 MPa and E_{cf} = 120,000 MPa, and GFRP tensile strength = 400 MPa and E_{gf} = 25,000 MPa. The CFRP and GFRP are assumed to rupture immediately after their tensile strength is reached. Addition of FRP to the underreinforced section increases its flexural capacity as expected. Although the maximum forces that can be provided by CFRP and GFRP are equal, their effects on the beam behavior are different due to the differences in their stiffnesses properties. CFRP-repaired beam has the largest flexural capacity (39% higher than that of the control beam) which corresponds to the rupture of CFRP sheet beyond which the beam converts to a regular steel reinforced concrete beam. A drop in the moment capacity follows and concrete crushing takes place at the same deformation as in the control beam. Depending on the amount of CFRP used, the drop in the moment capacity in some cases may be unacceptably steep. The GFRP-repaired beam on the other hand displays an increase in flexural capacity of about 22% with about 18% reduction in ductility. Crushing of concrete in this beam takes place before GFRP reaches its ultimate stress. The initial stiffnesses of the beams strengthened by the composites are not significantly different than that of the control beam.

3.3 Columns

Perhaps the most important and feasible technique to improve the flexural strength and ductility, and shear strength of a column is by confining the concrete. It is well established [12, 13, 14] that the closely spaced lateral reinforcement increases the compressive strength and the effective ultimate compressive strain of concrete. The lateral pressure responsible for enhancing strength and ductility of concrete can also be easily provided by ACM. In a test program the strength of 24 MPa concrete was enhanced to 190 MPa due to confinement by ACM [15]. Fig. 8 shows the stress-strain curves of concrete under compression confined by lateral and longitudinal steel arranged in different configurations. It is obvious that the detailing of reinforcement

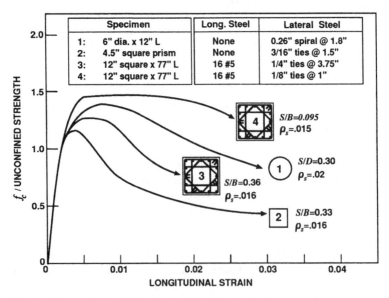

Fig. 8. Comparison of circular and rectilinear confinement

plays an important role in determining the concrete behavior. Fig. 9 shows the mechanisms of confinement in circular and rectangular configurations. In circular shapes, the entire area of concrete in the core at the level of lateral reinforcement is effectively confined whereas in the rectangular shapes, only part of the core area is effectively confined the extent of which depends on reinforcement configuration. Larger number of longitudinal bars supported by tie corners in rectangular cores results in a larger fraction of the core concrete effectively confined and hence a more efficient confinement mechanism. The area of the effectively confined concrete is further reduced at sections away from the lateral reinforcement level.

A popular technique to repair columns is jacketing which is commonly used in earthquake zones and which traditionally uses steel and concrete. Jacketing can also be accomplished by using ACM fibres, rods or tapes which can be used in the form spiral or hoops before curing of the epoxy, followed by shotcreting of the column (Fig. 10). ACM fabric can also be used to wrap the column followed by an injection or casting of grout between the column and the fiber wrap. Alternately, the grout can be cast first around the column followed by wrapping of ACM (Fig. 11). If expansive grout is used, the ACM wrapping will be prestressed causing an active confining pressure on the column resulting in a stiffer column behavior compared with that of a conventionally confined column. When ACM fabric is used to confine a column,

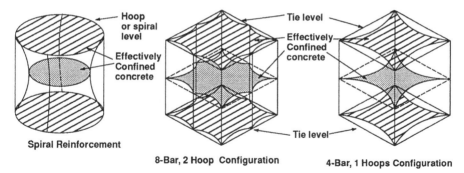

Fig. 9. Effectively confined concrete area

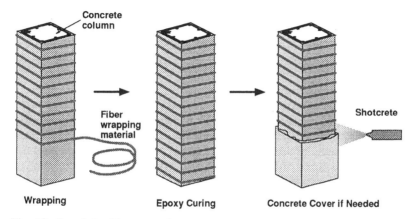

Fig. 10. Repair by fiber wrapping

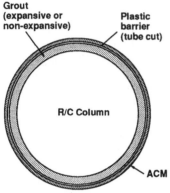

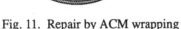

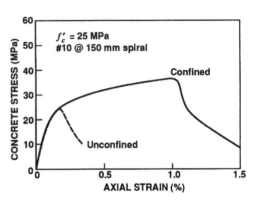

Fig. 11. Repair by ACM wrapping

Fig. 12. Comparison of ACM-wrapped confined concrete and unconfined concrete

there is no reduction in the effectively confined concrete area along the column length (Fig. 9). The efficiency of confinement is therefore better in this case compared to the columns in which reinforcement is used at discrete points. Glass fibres degrade significantly and lose tensile strength when they are subjected to alkali penetration for long period of time [7, 8]. ACM wraps containing glass fibres should not be used directly against the fresh grout. A plastic barrier between the glass fibre based ACM and the grout should work well. Care should be taken to eliminate wrinkles in the fabric wrap.

Demers [16] investigated the confinement effects of fibre reinforced plastic sheets wrapped around damaged and undamaged columns. Sixteen cylinderical column specimens - 300 x 1200 mm were tested in which the main variables were the thickness of FRP, strength of concrete, size of the longitudinal bars, and the spacing of the spirals. Eight columns were preloaded to their axial load carrying capacity before the application FRP, whereas the other eight were strengthened directly. The 0.33 mm thick CFRP tape used had the following characteristics in the direction parallel to the fibres: ϵ_u =0.015, f_u = 1140 MPa, E = 76000 MPa. The spacing of the circular steel hoops ranged between 150 mm and 300 mm. The increase in the axial load carrying capacity of the columns that can be attributed mostly to the ACM confinement ranged from 3.2% to 30.9%. The axial strain corresponding to the maximum stress of the concrete ranged from 0.0038 to 0.01. Fig. 12 illustrates the stress-strain behavior of the confined concrete in a typical ACM-wrapped specimen.

3.4 Column repair example
Consider a column shown in Fig. 13. The compressive strength of concrete is assumed to be 25 MPa and the yield strength of steel is 400 MPa. Without including any capacity reduction factors and ignoring the strain hardening of steel, the axial load carrying capacity of the column P_o is 13,863 kN at small deformations [3]. At large deformations when the cover is spalled off and confinement from spiral comes into effect, the capacity of the column is about 13,833 kN. The amount of lateral steel in the column is designed to compensate for the loss of cover concrete [3] considering that the enhancement in concrete strength due to confinement is 4.1 times the lateral pressure [12].

Table 2. Summary of results from column example

Column	Capacity at small deformations (kN)	Capacity at large deformations (kN)
Control	13 863	13 833
Damaged	10 989	≈11 000
CFRP-repaired	13 863	21 177-31 234
GFRP-repaired	13 863	18 661-31 234

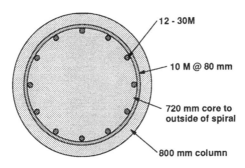

Fig. 13. Column cross-section

Now consider the column is damaged such that its cover concrete is completely spalled off, the 30M and 10M bars have lost the outer 1 mm thickness of steel due to corrosion and core concrete is delaminated such that its cross-sectional area is reduced by about 2%. The column in this state will display a capacity of 10,989 kN. A comparison of behavior of healthy control column with that of the damaged column is shown in Fig. 14. Due to the open spaces in the damaged core, the confinement stresses will not develop effectively at large deformations and the original column capacity will not be restored. The figure also shows the behavior of GFRP-repaired and CFRP-repaired columns. Same properties of CFRP and GFRP are used here as in the beam example described earlier. The radius of the repaired column is larger than the original diameter by the thickness of fibre wrap which is assumed to be 3 mm for CFRP and 6 mm for GFRP. In the repaired column at large deformations, the original core is confined by steel spiral as well as the FRP jacket. The area outside the original core is subjected to lateral pressure by FRP jacket only. The repair material/concrete is assumed to have $f'_c = 25$ MPa. Table 2 lists calculated capacities of all the columns. For each repaired column a range of capacities is shown for large deformations. The lower value is calculated corresponding to the yield strain of spiral steel and the upper value approximately corresponds to the rupture of FRP assuming that the spiral steel will continue to provide the lateral confining stress at yield level and concrete at such a large lateral strain will continue to resist enhanced stress. The most probable load-strain behaviour curves for these columns are shown in Fig. 14.

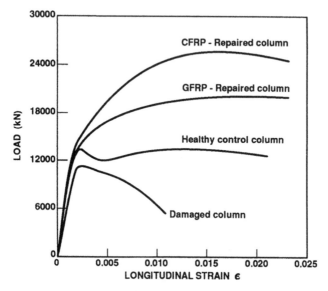

Fig. 14. Load deformation behavior of all of the columns

4 Current research

In a research program currently underway at the University of Toronto, ten half-scale models of the example column discussed above are under study for their short and long term behavior. The 400 mm nominal diameter and 1.52 m. long columns were cast with 25 MPa concrete to represent bridge columns along Ontario highways constructed a few decades ago. To simulate the field corrosion damage, eight columns were subjected to accelerated corrosion in the middle third of each column. A small tank was built around the middle third of each column which contained 2% chloride solution. The wall of the steel tank was used as an external cathode and all the six longitudinal bars were used as anodes. Fig. 15 shows a typical column with the tank and the corrosion damage observed in the middle third after the tank is removed. Further corrosion is continuing currently to cause further damage. The repair of the columns will be made on the lines of the scheme shown in Fig. 11. In addition to restoring the capacity of the column or even enhancing it beyond the original value, one of the main objectives of the study is to evaluate the effectiveness of ACM wraps in reducing the rate of corrosion in the columns if all the contaminated concrete is not removed but the ingress of water and air to the affected area is minimized. The contaminated concrete may be left in place inadvertently in some cases while in others intentionally because removal of this concrete may cause temporary structural weakness during the repair which may have serious consequences if the structure is to be kept operational.

Of the eight corroded columns, six will be repaired using three repair schemes. One control undamaged column, one damaged unrepaired column and three repaired

Fig. 15. Coroded test columns

columns will form one set of specimens in which the short term properties of specimens including structural behavior up to failure under concentric load will be studied. The second identical set of column specimens will be monitored for long term properties including the rate of corrosion as affected by the repair techniques.

5 Concluding remarks

This paper describes typical damage to concrete structures caused by environmental effects and evaluates the causes which can be corrected by proper design and construction engineering practices. The rehabilitation of concrete structural components using advanced composites is discussed. In addition to presenting results from the experimental work, examples for the repair of beams and columns are given in which carbon fibre reinforced plastics and glass fibre reinforced plastics are used. The probable behaviour of beams and columns is presented and the effects of varying the type of FRP is discussed. It is suggested that with appropriate repair design the ACM-rehabilitated members can regain the lost capacity and display satisfactory ductility performance. The strength gain can be very substantial in the case of columns due to confinement provided by ACM, particularly if expansive grout is used between ACM and the old concrete.

6 References

1. Dunker, K.F. and Rabbat, B.G. (1993) Why America's bridges are crumbling. Scientific American, 66-72.
2. Bickely, J.a., Neale, K.W. and Fabbruzzo, G. (1993) Market potential and identification of suppliers of advanced industrial materials for construction of bridges and other structures. Industry Science and Technology Canada.
3. Building code requirements for reinforced concrete (ACI 318-95) and commentary (ACI 318M-95), (1995) ACI Committee 318. American Concrete Institute, Detroit, 369 p.
4. Code for the design of concrete structures for buildings (CAN3-A23.3-94) (1994). Canadian Standards Association, Rexdale, 199 pp.
5. CEB-FIP Model code for concrete structures: (1978) CEB-FIP international recomendations, 3rd ed., Comite Euro-Internatial du Beton, Paris, 348 p..
6. Taylor, P. (1995) Durability of concrete - South African Portland Cement Institute suggested standards. Seminar at the University of Toronto.
7. Mufti, A.A., Erki, M-A, Jaeger, L.G. (1991) Advanced composite materials with application to bridges. Canadian Society for Civil Engineering, 297p.
8. Mufti, A.A., Erki, M-A, Jaeger, L.G. (1992) Advanced composite materials in bridges and structures in Japan. Canadian Society for Civil Engineering, 43p.
9. Laboissiere P. (1993) Current and future applications of advanced composite materials in structural engineering. Canadian Society for Civil Engineering, 43p.
10. Meier, U. and Kaiser, H. (1991) Strengthening of structures with CFRP laminates, Advanced Composite Materials in Civil Engineering Structures. Proceedings of the Speciality Conference, Las Vegas, Nevada, ASCE, pp.288-301.
11. Sharif A., Al.Sulaimani, G.J, Basunbul, I.A., Baluch, M.H. and Ghaleb, B.N. (1994) Strengthening of Initially loaded reinforced concrete beams using FRP plates. ACI Structural Journal, V. 91, No. 2, pp. 160-168.
12. Richart, F.E., Brandtzaeg, A. and Brown, R.I. (1929) The failure of plain and spirally reinforced concrete in compression. University of Illinois, Engineering Experimental Station, Bulletin No. 190.
13. Sheikh, S.A. and Uzumeri, S.M. (1980) Strength and ductility of tied concrete columns. Journal of Structural Division, ASCE, V. 106, No. 5, pp. 1079-1102.
14. Sheikh, S.A. and Toklucu, M. (1993) Reinforced concrete columns confined by circular spirals and hoops. ACI Structural Journal, V. 90, No. 5, pp. 542-553.
15. Norris, M.S. and Nanni, A. (1993) Construction and repair of concrete members with FRP reinforcement. Construction Technology Update, pp. 542-553.
16. Demers, M. (1994) Determination des parametres influencant le comportement des colonnes en beton confinees par une enveloppe mince en composite d'avant-garde. Master of Applied Science Thesis, Faculty of Applied Sciences, Department of Civil Engineering, University of Sherbrooke, Quebec, Canada, 98 p.

17 OFU BRIDGE REPAIR PROJECT

G. TJUGUM
Rescon AS, Sagstua, Norway

Abstract

This paper is a description of an ongoing practical repair research project on a Norwegian coast bridge financed by the Public Roads Administration, the company Rescon and the Norwegian Industrial and Regional Development Fund (SND).

The paper describe a way of working and give no results, because this will come later, as the corrosion process on a coast bridge is a long term process.

Keywords: Monitoring, concrete repair, corrosion.

1 Introduction

The OFU Bridge Repair Project is a rather unique project linking research and technical development (RTD) to the actual rehabilitation of the Gimsøystraumen Bridge in the Nordland County, Norway. The project consists of two major contracts dealing respectively with RTD and trail repairs. The project period is 1993-1996. The main partners in the project are the company Rescon and the Public Roads Administration.

The contract dealing with RTD is a Public Sector Research and Development Contract (OFU contract). The contract is financed by the Public Roads Administration, the company Rescon and the Norwegian Industrial and Regional Development Fund (SND), which administrates OFU contracts. The contract amounts to NOK 16,55 mill. and is part of the RTD Project TUBTU, (Technical Development of Bridges and Tunnels) of the Public Roads Administration Programme.

The other contract, including trial repairs of the bridge, is between the contractor Resconsult and the Nordland County Public Roads Administration. Rescon is the holding company of Resconsult. The repairs are performed only during the summer months and were started in 1993 and run through to 1995. The cost of these repairs

Integrated Design and Environmental Issues in Concrete Technology. Edited by K. Sakai. Published in 1996 by E & FN Spon, 2–6 Boundary Row, London, SE1 8HN, UK. ISBN 0 419 22180 8.

amounts to NOK 10 mill., which is financed by the bridge owner, Nordland County Public Roads Administration.

The RTD project (OFU contracts) contains the following sub-projects:

1. Foundation for RTD
2. Product Development
3. Product Documentation
4. Equipment
5. Quality Assurance
6. Information and Marketing
7. Doctoral Research Studies
8. Administration

Foundation for RTD includes detailed plans and specifications of actual trial repairs and specifications of actual trial repairs and evaluation of both repair methods and repair work. Another part of this sub-project is problem definition. This ensures that the company's product development work will meet the requirements of the Public Roads Administration as a bridge owner.

Central activities of the project are Product Development and Product Documentation. The latter is performed both by laboratories in Norway and abroad.

The Quality Assurance project, no. 5, is designed to implement activities in Rescon that will result in the company receiving ISO-9001 classification.

There is a clear need for a strong and determined RTD effort into the durability of concrete structures. These projects emphasize this by running two PhD degree programmes in the area. The doctoral studies are within the main aims of the whole project.

In order to achieve quality the Trial Repair Contract emphasizes the need to develop detailed work procedures and routines. Again, thorough and reliable documentation of all relevant factors is of the utmost importance. Because of the need for reliable documentation, installation of a monitoring system is a major part of the Trial Repair Contract.

The monitoring system is considered of great importance both in assessing the quality of various repair methods used, and in studying various hypotheses concerning the durability and repair of concrete structures.

2 Problem description

Deterioration of concrete structures is a world wide problem and it seems to be increasing with time. If all concrete structures were to be repaired in the next few years in this country, the cost would run into billions of NOK. In reality of course, repair activities will be on a much lower scale, but this means that many structures will deteriorate even more.

Because of the long coast line in Norway where concrete bridges are frequently exposed to a harsh marine environment, this deterioration is a major problem for the Public Roads Administration. The Nordland County is one of the regions in Norway with a large number of bridges in this environment and consequently one of the regions with many damaged bridges.

There is an urgent need to develop cost effective maintenance and repair methods, and this is a formidable task. The mechanisms of deterioration are both complex and many. Only concentrated RTD projects into durability problems may solve this. The problems are both complex and long-term ones. This complicates the RTD work and calls for expertise from many fields of specialisation.

In addition to the technical performance of repair methods, the actual work procedures present great challenges. Important factors are the sensitivity of various methods to work performance, organization and quality assurance. Development of cost effective maintenance and repair methods are therefore important aims of the project.

3 Project organization

The project is organized as a co-operation between Rescon and the Public Roads Administration. A total of 15 persons from the Public Roads Administration are involved part-time in the project. They are collected from both the Nordland County Roads Office, the Bridge Department and the Road Research Laboratory. Rescon involves about 20 persons, also part-time.

The project is run by a Project Manager coming from the Nordland County Roads Office. The Project Manager reports to a board of representatives from Rescon and the Public Roads Administration.

Two advisory groups are attached to the project, a reference group and a doctoral degree forum. Members of these two groups are appointed on the basis of having high competence in specialized areas. The doctoral degree forum includes PhD degree advisors from the University of Oslo and the University of Trondheim, and representatives from both the firm Rescon and from the Road Research Laboratory.

The project also has three working groups attached to the project. These are:

- Working Group for Trial Repairs
- Working Group for Problem Definition and Product Documentation
- Working Group for Instrumentation, Documentation and Verification (IDV)

These groups have representatives from Rescon, Public Roads Administration, research institutes/universities and from private consultants.

4 Gimsøystraumen Bridge

Gimsøystraumen Bridge is located 30 km west of Svolvær on the E 10 road in the Nordland County. The bridge is a link between two of the Lofoten Islands, Austvågøy and Gimsøy. The longitudinal bridge axis is roughly east-west. The climate is severe with strong winds and salt spray from the sea. The bridge length is 840 m and the environmental effects vary greatly along the bridge as the bridge deck level varies from 4 m to 30 m above sea level.

The bridge is a post-tensioned, free cantilevered, box construction. The box height varies from 2,2 m to 7,4 m. The designed concrete strength for substructures was C35 with 50 mm cover of rebars. For the bridge deck the corresponding values were C40 and 30 mm. Fixing bars were allowed in the cover zone in horizontal construction elements. The designed quality of reinforcement bars was Ks 50 and Ks 40 for stirrups.

The use of fixing bars within the cover zone has lead to the development of deterioration. The most serious areas of deterioration are close to the sea level. The bridge was constructed between 1979 - 1981 by Ing. T. Furuholmen as contractor and Dr.Ing. Aas-Jakobsen as consultant.

5 Trial Repair I, 1993

In June - September 1993 it was made 4 test areas (trial repair) of about 100 m^2. The areas were carefully diagnosed and monitored for continuous information gathering in coming years.

Trial area 1
Problem: Visible damage caused by reinforcement corrosion on fixing bars and stirrup. Chloride content at the reinforcement up to 2% of cement weight.

This trial area is divided into two smaller areas 1-A and 1-B.

Trial area 1-A

- Water chiseling into reinforcement.
- Sandblasting.
- Dry spraying of special mortar tailormade for dryshot. Minimum 40 mm reinforcement covering.
- Immediately spread by curing compound.
- After some weeks sandblasting.
- Coating with elastic cement latex based slurry 3,5 kg/m^3 was able to bridge moving cracks 0,2 mm even at - 20°C.

Trial area 1-B
As 1-A, but the special mortar was added to a special gas phase inhibitor.

Main purpose
To check the effect of traditional rehabilitation chloride infected concrete, about 2 cm behind the reinforcement was removed. This had to be stopped after a very small area, because the reinforcement bended out. The testing was immediately modified, and the concrete was just removed into the reinforcement, so the reinforcement will still be baded in chloride contaminated concrete. The test will compare effect of a high quality cement mortar with and without inhibitor on the corrosion development in reinforcement. The inhibitor is supposed to migrate to the reinforcement and cover and protect it. The top layer will cover and bridge even moving cracks and give a Cl⁻ and CO_2 tight barrier protection.

Trial area 2
Problem: Few visible surface damages. Low corrosion. Chloride concentration at reinforcement 0,5-0,8% of cement weight.

- Sandblasting
- Traditional repair of small wounds and cracks.
- Impregnation with two layer of penetrating and moving inhibitor.
- After 2-3 days coated with 3,5 kg cement latex based elastic slurry with adhesion of gas phase inhibitor.

Main purpose
Making an efficient elastic cement latex based crackbridging barrier against Cl^- and CO_2. See if migrating gas phase inhibitor can migrate through the concrete covering and stop or delay corrosion of reinforcement.

Trial area 3
Problem: No visible damage. Partly corrosion on reinforcement. Cl^- concentration at the steel level - 0,7% of the cement weight.

- Sandblasting
- Repair of small damage and cracks in traditional way.
- Coating with elastic cement latex based slurry 3,5 kg/m^2.

Main purpose
See the effect of an effective crackbridging barrier cement slurry on corrosion development.

Trial area 4
Problem: Very few visible damages. Active corrosion situated Cl^- concentration at steel level 0,4-1,2% of cement weight.

- Sandblasting
- Repair of small damages in traditional ways.
- Hydrophobating the concrete surface with a silane based.

Main purpose
Study the effect of hydrophobation in «wet» field conditions. Corrosion, distribution and penetration of Cl^- and CO_2.

6 Trial Repair II, 1994

This trial repair was done on the bridge pillars, June - August 1994. Corrosion situation is not serious for any of the pillars, so after careful consideration one chose to give them preventive surface treatments. Test areas are bigger than in 1993 and goes from foundation up to the bridge.

Trial area 1
Hydrophobating with silan 0,4 kg/m^2.

Main purpose
As reference for good Cl$^-$-barrier properties.

Trial area 2
Hydrophobating and coating with diffusion open Cl$^-$ and CO_2-barrier paint 0,2 kg/m^2.

Main purpose
Together with trial area 1 to study the effect with and without CO_2-barrier.

Trial area 3
Elastic cement latex based slurry on top of hydrophobated concrete crack-bridging effective barrier slurry against Cl$^-$ and CO_2.

Main purpose
To gain experience with hydrophobating as a primer under elastic cement slurry.

Trial area 4
Cement based putty for evening and pore filling the concrete surface 2-3 kg/m^2. Coated on the top with a diffusion tight, flexible, crackbridging epoxy 2-3 kg/m^2.

Main purpose
To study the efficiency for O_2, Cl$^-$ and CO_2 tight barrier membrane on corrosion development.

7 Trial Repair III - 1995

At the time of writing this trial repair III is not started, but the principle will be traditional repair of small concrete wounds and so different coating depending on covering thickness and chloride content. There will also be poured different concrete elements and coated with 6-7 different material combination and fixed on the bridge for following up.

8 Corrosion monitoring

Installation of sensors and probes for the registration of

- corrosion state (steel reinforcement)
- factors affecting the corrosion state

Corrosion monitoring - why?

Goals
I.   To evaluate how concrete rehabilitation affects both the severity of corrosion and the factors affecting the corrosion state.
II.   To follow the long-term corrosion development in a concrete structure.
III.   To gain experience with corrosion monitoring of concrete structures.

The corrosion monitoring program at Gimsøystraumen bridge is focused on comparing

Rehabilitation areas   (Repair areas)
with
Untreated neighbour areas   (Reference areas)

Instrumentation techniques for the monitoring of the corrosion state

Corrosion of steel in concrete is an electrochemical process which can be detected by electrochemical measurement techniques.

At Gimsøystraumen bridge the corrosion state of the steel reinforcement is determined by

- Corrosion potential
- Linear polarisation resistance

Corrosion potential (Volts)

The potential of the steel gives a rough indication of the severity of corrosion.

Low potential   ⇒   High corrosion activity
High potential   ⇒   Low corrosion activity

The measurement is carried out by using a reference electrode and a volt meter.
The corrosion potential is given in volts with respect to the reference electrode.
At Gimsøystraumen bridge four types of reference electrodes are installed:

- Manganese dioxide
- Silver/silver chloride
- Graphite
- Lead

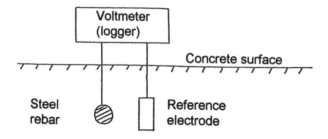

Fig. 1.   Measurement of rebar corrosion potential using an embedded reference electrode.

Linear polarisation resistance, LPR (Ohm or Ohm*m²)

The electrical response of the steel to an impressed electrical signal (voltage or current) near the corrosion potential is indicative of the magnitude of the corrosion current.

The ratio $\dfrac{\text{Change in potential (V)}}{\text{Change in current (A or A/m}^2)}$

is called   the linear polarisation resistance, LPR (ohm or ohm*m²)

High LPR-value $\Rightarrow$ Low corrosion current (rate)
Low LPR-value $\Rightarrow$ High corrosion current (rate)

If the measured steel area is known the corrosion rate ($\mu A/cm^2$) can be calculated by using a proportionality factor called the «LPR constant».

Notice that the chloride induced rebar corrosion is a localized type of attack and that the average corrosion rate determined by the LPR technique may underestimate high local corrosion.

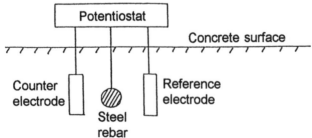

Fig. 2   LPR measurement in concrete by using embedded electrodes (schematic).

During Trial Repair II (1993) the rebars intended for LPR measurements were cut in 8 cm length in order to have defined steel areas. This was not done during Trial Repair II (1994) and Trial Repair III (1995).

FCB - Probe

In mechanically repaired areas where heavily chloride contaminated concrete was removed by water jet, corrosion probes (named FCB probe) have been installed near the rebars. The probe, consisting of a reference electrode and steel rods in depths 20, 30 and 40 mm below the new shotcrete surface is used for potential and LPR measurements. By monitoring the rebar and rod potentials and the LPR of the rods new chloride ingress will be detected.

Instrumentation techniques for the monitoring of factors affecting the corrosion state

The monitoring program at Gimstøystraumen bridge also includes probes for the determination of factors affecting the corrosion state, i.e. probes for the measurements of:

- Oxygen limiting current in the concrete
- Electrical resistance in the concrete
- Relative humidity in the concrete
- Concrete temperature

Oxygen limiting current is an electrochemical parameter; the other parameters are purely physical.

Oxygen limiting current (A/m^2)

The corrosion process can be divided into two electrochemical reactions:

- Oxidation of iron $\quad Fe \rightarrow Fe^{2+} \quad$ (Anodic reaction)
- Reduction of oxygen $\quad O_2 \rightarrow OH^- \quad$ (Cathodic reaction)

The rate of the cathodic reaction must equal that of the anodic one. Thus the rate of the oxygen transport to the steel surface sets an upper limit to how fast the anodic reaction (the dissolution of iron) can proceed. Maximum oxygen conversion is determined by cathodic polarisation of a steel rod to a potential where oxygen reduction is the only reaction occurring. When further polarisation (to more negative potentials) does not result in increasing current the limiting current is reached. This current is determined only by how fast oxygen reaches the steel surface by diffusion.

Notice: This is an ideal picture of what is going on during the measurement. Several factors may affect the result and the technique is not yet fully developed.

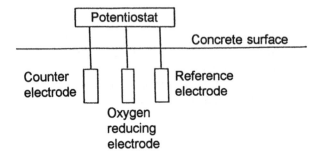

Fig. 3.   Measurement of oxygen limiting current in concrete by using embedded electrodes (schematic).

Electrical resistance (Ohm)

The flow of electric currrent in the concrete plays an important role in the corrosion process. Low electrical resistance stimulates corrosion. High resistance reduces corrosion. Several factors affect the electrical resistance:

- Concrete/pore structure
- Water content
- Pore water composition
- Temperature

At Gimsøystraumen bridge the following probes have been installed:

- **Multiring electrode**
  Measurement of the resistance between nine embedded steel rings placed in gradually increasing depths below the concrete surface.
- **Steel rods**
  Measurement of the resistance between two embedded steel rods placed a few cm apart.
- **Conductive coating**
  Measurement of resistance between two stripes of a conductive coating placed a few cm apart on the concrete surface.

The measurements are carried our with AC-voltage.

Relative humidity (%)

The electrical resistance is highly dependent on the moisture content in the concrete. As a supplement to these measurements probes have been installed for the measurement of relative humidity inside the concrete.

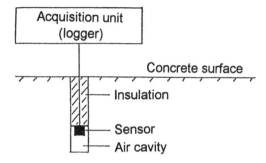

Fig. 4. Measurement of relative humidity in concete (schematic).

Temperature (mV → °C)

All the parameters described above are sensistive to temperature. Therefore, probes for the registration of temperature (thermocouples) have been installed near the other probes.

The location of sensors and probes

The sensors and probes are arranged in groups (one group consisting of different sensors/probes) in the repair and reference areas in the hollow sections of the spans (several meters above sea level) and in columns (a few meters above sea level).

The following bridge areas have been selected for instrumentation:

Trial repair I (1993)	Hollow section of span 2
	Wall (south)
	Bottom
	Wall (north)
Trial Repair II (1994)	Columns # 3, 4 and 5
	Wall (east)
	Wall (west)
Trial Repair III (1995)	Hollow section of span 1
	Wall (south)
	Wall (north)

The sensors and probes have been, whenever possible, installed in drilled holes from the inside of the concrete walls. In some locations the installation has been done from the outside (see Figure 5). The drilled holes where filled with a cement mortar without polymer additions.

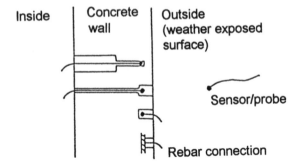

Fig. 5   Illustration showing sensor installation at Gimsøystraumen bridge (schematic).

The exact location of each group of sensors and probes within the repair and reference areas was selected after an evaluation of the results from a preceding corrosion state survey.

There were three «levels» in the selection procedure:

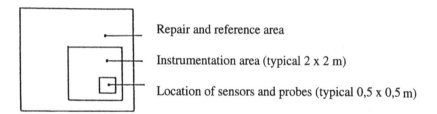

The instrumentation area, selected after a general condition survey, was investigated in detail:

- Rebar pattern and concrete cover (Cover meter)
- Corrosion potential mapping (Bloodhound)
- Corrosion rate (Gecor 6)
- The condition of concrete surface and rebars (Visual inspection)
- Chloride profiles (samples for analysis in laboratory)

Sensors and probes were then located in the area with highest probability of corrosion.

Data aquisition

Monitoring cables from sensors and probes have been led inside the hollow sections of spans and columns and connected in junction boxes, one box for each group of sensors/probes (a total of 23 junction boxes). The junction boxes are connected to a monitoring device consisting of a logger and a system for remote control and data acquisition (by modem).

The system has logged data four times a day since January 1994.

Table 1 Sensors and probes installed at Gimsøystraumen bridge 1993-1995.

Type of instrumentation		Number of sensors and probes			
		Trial Repair I (1993)	Trial Repair II (1994)	Trial Repair III (1995)	Total
Corrosion pot.	Manganese	15	12	4	32
	Ag/AgCl	--	6	--	6
	Graphite	11	--	--	11
	Lead	11	--	--	11
Polarisation resistance, LPR		12	12	4	28
FCB-probe		2	--	--	2
Oxygen lim. current		15	6	2	23
Resistance	Multiring	14	--	--	14
	Rods (2)	14	--	--	14
	Rods (4)	--	12	4	16
	Coating	7	--	--	7
Relative humidity		8	8	--	16
Temperature		16	18	4	38

Two types of measurements are not carried out by the remote monitoring system, i.e.

- All LPR measurements
- Measurements of oxygen limiting current on probes installed in 1993. (Probes installed in 1994 and 1995 are.connected to the remote monitoring system).

9 Conclusion

Our experience is that this way of working ought to have a good chance to give good results, but technical approved results will first be available in 5 - 10 years during continuous monitoring of the corrosion process.

18 THE DESIGN OF CONCRETE STRUCTURES AGAINST UNFAMILIAR LOADING CONDITIONS

N. VITHARANA
Major Works Division, Department of Water Resources, NSW, Australia
K. SAKAI
Hokkaido Development Bureau, Sapporo, Japan

Abstract
The knowledge on the design of concrete structures against dead and live loads is relatively well-established compared with that against strain-induced loadings. The stresses due to the strain-induced loadings could be much more severe than those imposed by the applied loadings. Therefore, the evaluation of strain-induced loadings should be based on rational approaches in order to achieve economical and durable structure designs. Unfortunately, the available design aids or recommendations are very limited in this regard. This paper reviews the current approaches adopted and also presents some of the significant aspects of the behaviour of concrete structures subject to strain-induced loadings.
Keywords: Concrete structures, design, hydration, shrinkage, temperature, unfamiliar loadings.

1 Introduction

The imposed loadings are usually specified for a particular structure under different conditions, and accordingly the designers are required to design the structure to resist the imposed loadings. This is the case with many design practices or standards in which the data are provided and recommendations are usually made based on the generally-accepted criteria. If a designer is required to consider these aspects from the fundamentals, the design process would be time-consuming as well as possibly influenced by the personnel preferences or experiences. Furthermore, the incorporation of statistical variations would be quite impossible and very expensive in a design office environment. However, the blind adherence to almost all the

Integrated Design and Environmental Issues in Concrete Technology. Edited by K. Sakai. Published in 1996 by E & FN Spon, 2–6 Boundary Row, London, SE1 8HN, UK. ISBN 0 419 22180 8.

provisions and recommendations given in a design practice, particularly in uncommon situations or beyond the limitations of the applicability of the design provisions, is a step beyond innovation or suppression of new ideas or a risky affair when the consequent failures would be the loss of human lives or the waste of public money.

Even when the loadings are given, the majority of current design philosophies are self-contradictory, eg, the structural analysis is elastic (linear) whereas the structure behaviour is non-linear in the design of concrete structures particularly for ultimate limit state of strength. A broad allowance such as 30% of load redistribution is usually recommended, irrespective of the actual situation. This does not mean to say that those structures, designed based on the conventional procedures, would collapse or need to be replaced immediately once a non-linear analysis methodology has been developed. Nevertheless, the time has now come to assess the validity of some of the conventional methods and also to observe the significance of these discrepancies in a rational way.

In the last decade or two, the design of concrete structures has become more versatile and challenging due to the introduction of different loading conditions [1,2,3] which were considered to be insignificant in the past either due to the lack of understanding or simply transferring the observed failures (eg, cracking, excessive deformations) to other causes. At present, in several standards, concrete structures are required to be rationally designed against unfamiliar loading conditions such as temperature, shrinkage and swelling [1,2,3,4]. The term *unfamiliar* is used herein mainly because the rational design against these loadings is not well unknown, but not the existence of such volumetric changes. This is quite unfortunate considering the fact that the strain is something which we can see physically but not the stress which is only measurable. These are usually known as strain-induced loadings, and the design & analysis processes are much more complicated than those with the mechanically-applied loadings. As a result, particularly in the past, rules of thumb approaches have been adopted such as minimum reinforcement ratios or the provision of contraction and expansion joints or arbitrary cooling at worst. However, in several modern codes of practice [1,2,3], the design provisions have been incorporated requiring the designers to calculate the strain-induced stresses in a rational manner. The history of the adopted temperature gradients for concrete bridge beams in the Australian practice is shown in Fig.1. The need has arisen after the early deterioration of many reinforced and prestressed concrete structures [eg, 5,6]. However, it is quite surprising that the highly-regarded North-American practice has ignored these aspects despite detrimental effects caused by them [5].

Although the above codes of practice are to be appreciated for the pioneering move, unfortunately the majority of the recommendations are irrational [7] and sometimes do not meet the basic principles of structural mechanics. Strain-induced loading conditions depend significantly on the exposure and ambient conditions. However, this has not been understood by many designers, eg, the temperature gradients developed for NZ, where the climatic are moderate, have been cursorily adopted for the Australian conditions as well as the conditions in Alaska.

2 Mechanism of strain-induced loading

The stresses due to induced strains are produced by two basic actions:

(1) Due to non-uniform strain gradients, eg, in beam members which deform according to the plane sections remain plane hypotheses, or in circular columns in the transverse direction. These are usually known as primary stresses.

(2) Due to structural indeterminacy or external restraints. This is due to the fact that the free end deformations of the members are restrained. This condition occurs in plane frames, multi-span continuous bridges, circular concrete tanks etc. The induced stresses are known as secondary stresses.

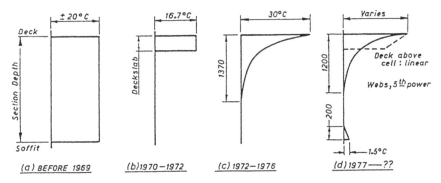

Fig. 1. History of design temperature gradients for the Australian conditions.

The most important aspect of the strain-induced loading is their dependency on the section and member stiffness. The induced loading is directly proportional to the stiffness, Fig. 2. As can be seen, the incremental loading is related to the tangential stiffness. The non-linear behaviour relaxes the strain-induced loadings significantly, and its negligence would result in a gross overestimate of the induced loading. This is very important in the case of reinforced concrete members where cracking is inevitably occurs, eg, the allowable stress under service loading is about 150 MPa, and the corresponding concrete strain thus far exceeds the cracking strain of concrete. Another important thing to note is the negligible value of the strain-induced loading at ultimate limit state due to the plastic deformations caused by the already applied loading. Therefore, in general, the strain-induced loadings are significant at service conditions only. Accordingly, in the design of concrete structures, where the serviceability limit states such as limited crack width govern, the strain-induced loadings should be incorporated. The structures coming under this category are fully & partially-prestressed concrete structures, water-retaining structures, containment structures, and structures exposed to aggressive environments etc.

The use of constant stiffness reduction factors to allow for load relaxation in reinforced concrete structures is quite invalid as the stiffness of concrete structures varies with the degree of loading due to the stiffness enhancement by the tension-stiffening effect. Therefore, the effective member stiffness due to tension stiffening is much higher than that calculated at a cracked section.

Figure 3 shows the significant of cracking in the predicted hoop forces and moments on a typical reinforced concrete cylindrical reservoir wall, compared with those based on code recommendations and also on uncracked section. In Figure 3, C_t represents the tensile strength of concrete f_t by the relationship: $f_t = C_t \sqrt{f_c'}$ (MPa) where f_c' is the concrete compressive strength.

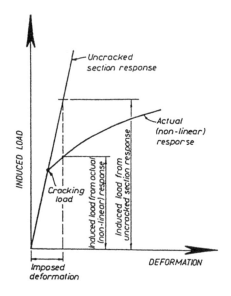

Fig. 2. Effect of cracking on the strain-induced load.

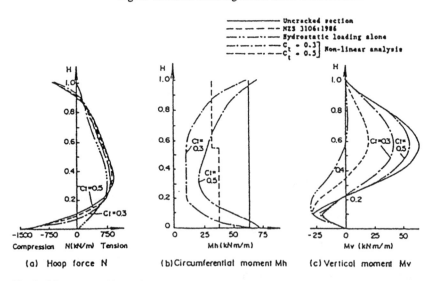

Fig. 3. Moments and hoop force for a typical reservoir wall under "summer" conditions.

3 Early-age stress generation

Hardening process of cement is exothermic. The amount and the rate of heat generation depends on the cement time, process temperature, environmental conditions, concrete composition etc. Although the significance of the heat of hydration on the generation of stresses has been recognised since the days of invention of cement, the methods adopted by many of the current design practices are quite irrational particular at a time when new materials, construction methods etc. are being innovated [8,9]. It should be noted that the cracking of concrete is a random process due to its dependency on many independent variables as well as interactive parameters. Therefore, in order to assess the tendency to cracking of concrete members, the probabilistic nature of cracking should be correlated with the theoretical predictions based on practical observations. The mechanism of early-age cracking under fully-restrained conditions is schematically shown in Fig. 4.

The cursory use of conventional recommendations could lead to serious errors. Such a case is the use of equivalent thickness for formwork or insulation in concrete placements in order to avoid a multi-layer heat-transfer analysis. This equivalence has been based on the assumption that there is no external input of thermal energy on the surface, eg, no solar energy. If this equivalent thickness is used in the presence of solar energy, the resulting errors are very significant [8].

The adoption of the adiabatic hydration models, which are easier to measure in practice, is incorrect when the actual reaction environment is not adiabatic. The adiabatic case would be closely present in thicker sections or sections with boundaries of low thermal diffusivity. This is due to the fact that hydration process is thermally-activated so are many other chemical reactions. But this effect has been ignored in many conventional heat-transfer analyses [9]. A simple non-adiabatic model is the Rastrup model [10] which considers the temperature/time history. A simple model based on a family of adiabatic curves has been recently proposed [8] considering the accumulated heat generation and the current process temperature indirectly. The error caused by the use of an adiabatic model for a 200 mm and a 600 mm wall section is shown in Fig. 5.

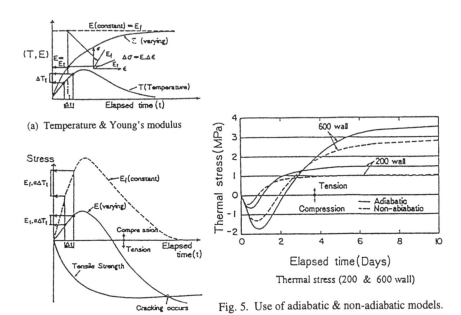

(a) Temperature & Young's modulus

(b) Stress & strength

Fig. 4. Early-age stress development.

Thermal stress (200 & 600 wall)

Fig. 5. Use of adiabatic & non-adiabatic models.

The temperatures generated on a 1000 mm dia column [11] by different cement types are shown in Fig. 6. As can be seen, HSC tends to generate more heat at a rapid rate compared with NSC. The availability of heat of hydration models for different cement types is therefore essential for the routine determination of the tendency to cracking. However, it should be noted that the evaluation of temperature is not of eventual importance as the tendency to cracking is governed by the ratio of the induced stress to the available tensile strength. Therefore, it may be that the HSC may have a less tendency to cracking compared with NSC. The stress ratio developed in a 1000 mm dia concrete column of 35 MPa and 100 MPa is shown in Fig. 7. As shown, there is an insignificant difference between HSC and NSC with respect to the tendency to cracking caused by heat of hydration.

Another example [12] is shown in Fig. 8 to highlight the fact that the conventional recommendations [9] for lowering the hydration-induced temperature development to reduce cracking in large foundations are quite unnecessary. Figure 8 compares the stresses induced in a multi-layer and single placements of concrete for thick foundation slabs of different thickness. It has been recommended [9], as well as generally believed, that concrete placements in multi-stages (or multi-layers) would reduce the tendency to cracking. This is based on the assumption that heat would rapidly dissipate in multi-layer placements due to shorter heat-flow path, and consequently lower temperature development. However, the fact which has been ignored is that the thermal stresses are induced due to the restraining effects. This is the case of a multi-layer placement for a large foundation pad, Fig. 8. In a multi-layer placement, the concrete at the bottom layers would be in tension of the same order as the top layer in a single-placement. It is quite encouraging to see that a recently constructed foundation slab (4.6 m thick) in Sydney was placed in a single layer (after a greater effort to convince), and no cracking was observed. This saved construction time and cost as well as the installation of concrete cooling methods.

4   Shrinkage and swelling-induced stresses

The shrinkage of concrete is caused by three basic actions: Drying, autogenous and carbonate. These are not discussed in detail here. Drying shrinkage occurs when the moisture is lost from the concrete during drying while autogenous (or self-desiccating) shrinkage occurs at early age when the moisture is consumed during hydration. The later is very critical [8] when low w/c ratios are used (say below 0.4) as in the production of high-strength concrete.

Nobody other than Bazant has carried out extensive research on the shrinkage characteristics. However, it would be very difficult to understand how the limited experimental data on 10 mm thick concrete mortar samples could be extrapolated to site concrete made with 40 mm coarse aggregate and cured in the ambient rather than in a electronically-controlled oven. Another aspect to note is that drying (diffusion) is a slow process and therefore how to separate out material shrinkage and structural shrinkage. Structural shrinkage is accompanied by simultaneous micro-cracking due to the highly non-linear moisture gradient in a given member. Also the cracking will occur instantaneously as soon as the surface of the member is allowed to dry as a free shrinkage strain of about 530 (micro strains) is imposed on the surface, compared with 100 (micro strain) corresponding to the cracking of concrete. Furthermore, the

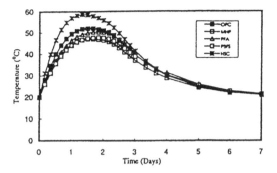

Fig. 6. Temperature generation by different cements in a 1000 mm dia column.

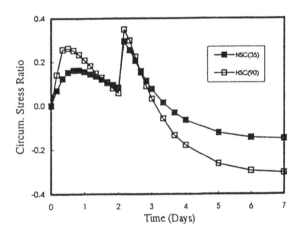

Fig. 7. Predicted stress ratios for HSC(100) & NSC(35).

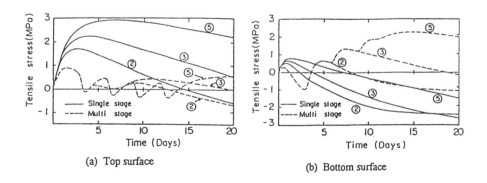

Fig. 8. Stress development in multi & single-layer concreting for a foundation pad.

formulations describing drying process in terms of pore humidity rather than the traditional moisture content is based on major simplifying assumptions on the isotherms (relationship between moisture content & pore humidity). Particularly, linear relationships are assumed when the isotherms are actually highly non-linear in some regions. According to the author's knowledge, only or two isotherms are reported in the literature.

Swelling occurs when a concrete member is allowed to absorb water. Swelling or shrinkage regain has been observed to be much faster than drying. The accurate prediction of swelling-induced loadings is very important [13] for some structures. Swelling-induced stresses would help to counteract short-term stresses caused by solar radiation and ambient temperature on a concrete reservoir wall. This would also help to understand why extensive cracking has not been observed in concrete reservoir walls although the calculated thermal stresses indicated severe cracking. This counteracting effect would save a quite lot of additional prestressing (effective prestressing above 8 MPa on reservoir walls would escalate the cost significantly). The loadings on a prototype reservoir wall is shown in Fig. 9.

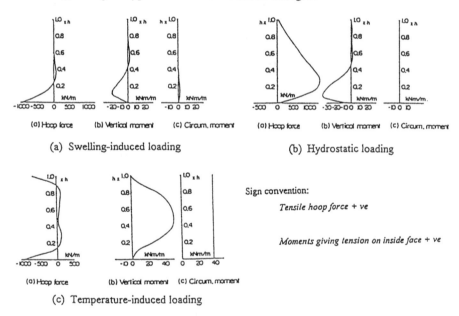

Fig. 9. Typical loadings on a cylindrical prestressed reservoir wall
(*height = 10 m, dia = 35 m, wall thickness = 225 mm*)

Another important aspect is the autogenous shrinkage occurring in concrete made of low w/c ratios. This has resulted in the early cracking and replacement of major works such as the Kinzua dam stilling basin in USA. Very few experimental data is available in this regard. Unlike drying shrinkage, autogenous shrinkage occurs at early age (in conjunction with hydration temperature) throughout the member thickness. If the member is axially restrained, through-depth cracking will occur. Through-depth cracking is quite undesirable due to its effect on the stiffness as well as the water-tightness. In a recent study [8], the available data on autogenous shrinkage was reviewed and some theoretical predictions were made. Some of these results are shown in Fig. 10. As can be seen, the autogenous shrinkage alone is severe enough to cause early-age cracking. Research on prototype restrained members made with different concrete compositions is urgently required.

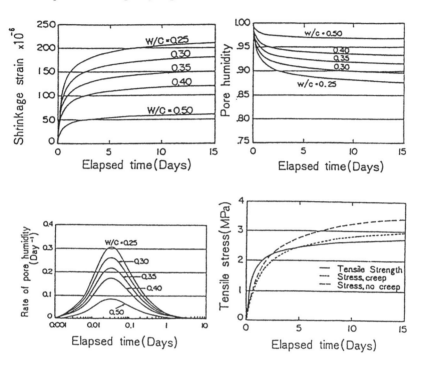

Fig. 10. Autogenous shrinkage in concrete members.

5   Conclusions

The consideration of strain-induced loadings is very important as recognised by several modern standards and design practices. Unfortunately, the significance of these loadings has not been recognised by some of the major standards and design practices despite major failures attributed to the negligence or the cursory adoptions, confirmed by the research findings. It is strongly recommended that these effects be considered rationally in the future particularly with the introduction of new cement types, concrete mix proportions, construction methods etc.

6   References

1.   Standards Australia (1983) *AUSTROADS Bridge Design Code*, Sydney.
2.   Standards Association of New Zealand (1986) *NZS 3106 Code of Practice for Concrete Structures for the Storage of Liquids*, Wellington.
3.   Standards Australia (1989) *AS 1170.1 SAA Loading Code*, Sydney.

4. PCI Committee Report (1987) Recommended Practice for prestressed concrete circular tanks, *PCI Journal*, July-Aug 1987, pp.81-125.

5. Slater, W.M. (1985) Concrete water tanks in Ontario, *Canadian Journal of Civil Engineering*, Vol. 12, pp.325-33

6. Holland, T.C., Ktysa, A. et al (1991) Use of silica-fume concrete to repair abrasion-erosion damage in the Kinzua Dam stilling basin, *American Concrete Institute, SP 91-40*, pp.841-63.

7. Vitharana, N.D., Priestley, M.J.N. and Dean, J.A. (1991) *Strain-Induced Stressing of Concrete Storage Tanks*, PhD Thesis Report 91-8, University of Canterbury, New Zealand.

8. Vitharana, N.D.(1994) Evaluation of Early-Age Behaviour of Concrete Wall Sections: Hydration & Shrinkage Effects, *Report No. 94/1*, Civil Engineering Research Institute, Sapporo, Japan.

9. ACI Committee 207. (1970) Mass concrete for dams and other massive structures, *ACI Journal*, Apr. 1970, pp.273-309.

10. Rastrup, E. (1954) Heat of hydration in concrete, *Magazine of Concrete Research*, Vol. 6, No. 17, pp.79-92.

11. Vitharana, N.D. (1995) Prediction of temperature and stress developments in high-strength concrete columns under heat-of-hydration effects, *Concrete'95, FIP International Conference*, Brisbane, Australia, pp.211-22.

12. Vitharana, N.D. and Sakai, K. (1995) Single and multi-stage concrete placements for large foundations: hydration-induced thermal stresses, *EASEC-5*, Gold Cost, Australia, pp.2311-15.

13. Vitharana, N.D. (1995) Rational evaluation of shrinkage and swelling-induced loadings in cylindrical concrete reservoir walls, *EASEC-5*, Gold Coast, Australia, pp.963-68.

19 DESIGN METHOD OF REINFORCED CONCRETE STRUCTURES ON THE BASIS OF DAMAGE

M. SUZUKI
Public Works Research Institute, Ministry of Construction,
Tsukuba, Japan
Y. OZAKA
Tohoku Gakuin University, Tagajou, Japan

Abstract
This study proposes a new method of evaluating a damage index for reinforced concrete members which suffer flexural failure after flexural yielding. It is obtained by comparing the accumulated energy absorbed by a member with the maximum absorbed energy considered to be the specific value for the member. Available experimental data were used to verify the damage index, confirming that it is very precise. The proposed damage index has been applied to seismic design as a measure for assessing the reliability of reinforced concrete structures.
Keywords: Accumulated absorbed energy, damage index, maximum absorbed energy, probabilistic approach, reinforced concrete structure, seismic design.

1 Introduction

In the future, seismic evaluations of structures will have to include an evaluation of their lifetime safety (lifetime risk) in addition to their safety when subjected to one or two specified earthquake loads. Such an evaluation method will have to include a means of studying safety as damage accumulates, and must include a rational evaluation of the degree of allowable damage to a member or structure, the amount of damage inflicted by an earthquake, and the effect on the structure and members of the ground and other related conditions.

When attempting to evaluate a damage index for a structure, it is advisable to compare the accumulated energy absorbtion with the critical amount of energy the structure can absorb.

On this understanding, we propose a method of evaluating a damage index for reinforced concrete members which suffer flexural failure after flexural yielding by

Integrated Design and Environmental Issues in Concrete Technology. Edited by K. Sakai. Published in 1996 by E & FN Spon, 2–6 Boundary Row, London, SE1 8HN, UK. ISBN 0 419 22180 8.

treating the accumulated absorbed energy as an index. The proposed damage index has been used as a measure of the reliability of reinforced concrete structures in a probabilistic approach, while at the same time an attempt was made to apply it to seismic design as a contribution to the establishment of a life−time risk evaluation method for reinforced concrete structures.

2 Definition of the damage index

In this study, the damage index D representing the degree of damage to a reinforced concrete structure is defined by the following formula:

$$D=\frac{W_{acc}}{W_{max}} \tag{1}$$

where,

W_{acc}: Accumulated absorbed energy, W_{max}: Maximum absorbed energy

D represents the degree of damage accumulated in a reinforced concrete structure due to earthquake motion from the time of construction.

3 Evaluation of the maximum absorbed energy

3.1 Maximum absorbed energy in the cross section $W_{c.max}$

3.1.1 Characteristics of energy absorption

Koyanagi et al. [1] have reported that 70% or more of the energy absorbed by a reinforced concrete beam that suffers flexural failure is absorbed by the tensile reinforcement. Their analysis indicates that the closer the tensile reinforcement ratio is to the compressive reinforcement ratio and the smaller the working axial compressive force, the higher the proportion of energy absorbed by the tensile reinforcement. Since this study deals with reinforced concrete members subjected to cyclic loading, almost all the energy absorbed by the section is taken up by the longitudinal reinforcement, and the maximum absorbed energy in the section $W_{c.max}$ is as follows.

$$W_{c,max}=S_s(\varepsilon_{su})\cdot(A_s+A'_s) \tag{2}$$

where,

ε_{su}: Ultimate strain of the reinforcement, $S_s(\varepsilon_{su})$: Area of the stress−strain curve of the reinforcement ($\sigma-\varepsilon$ curve) up to the ultimate state, A_s, A'_s: Sectional area of the tensile and compressive reinforcement

The ultimate state of the reinforcement corresponds to the moment when the member reaches the ultimate state.

3.1.2 Modeling the material characteristics

The $\sigma-\varepsilon$ curve of the reinforcement is assumed to be bi−linear.

Since using the $\sigma-\varepsilon$ curve for plain concrete leads to under−evaluation, the $\sigma-\varepsilon$ characteristics of confined concrete proposed by Muguruma et al. [2] were taken as the $\sigma - \varepsilon$ curve for the concrete.

3.1.3 Definition of the ultimate limit state

The proposal of Suzuki et al. [3] for ultimate limit state for the equilibrium of forces in the section was used as the flexural ultimate limit state. Because of the decrease in the resultant force of the compressive zone of concrete caused by strain softening, this ultimate limit state is the point where the strain in the tensile reinforcement stops rising and begins to fall, and is defined as the point where the bending moment in the bending moment–curvature ($M-\phi$) relationship abruptly declines.

3.1.4 Evaluating the ultimate strain ε_{su} of the longitudinal reinforcement

The ultimate strain ε_{su} of the tensioned reinforcement at the ultimate flexural state is evaluated using the ultimate limit point explained in 3.1.3.

From the equilibrium of forces in the section (See Figure 1),

$$T=C'+T' \tag{3}$$
$$T=A_s \sigma_s=pbd\,\sigma_s \tag{4}$$
$$C'=\frac{S_c(\varepsilon'_c)}{\varepsilon'_c}bx \tag{5}$$
$$T'=A'_s \sigma'_s=p'bd\,\sigma'_s \tag{6}$$

where,

T, T': Resultant force acting on the tensional and compressional reinforcement, C': Resultant force in the compressive zone of the concrete, σ_s, σ'_s: Stress of the tensional and compressional reinforcement, b: Width of the section, d: Effective depth, ε'_c: Concrete strain, $S_c(\varepsilon'_c)$: Area of the $\sigma - \varepsilon$ curve of the concrete in the ultimate state, x: Distance from the fiber under most extreme compression to the neutral axis, p, p': Tensional and compressional reinforcement ratios

Consequently,

$$pd\,\sigma_s=\frac{S_c(\varepsilon'_c)}{\varepsilon'_c}x+p'd\,\sigma'_s \tag{7}$$

By applying the Bernoulli–Euler hypothesis to the section,

$$x=\frac{\varepsilon'_c}{\varepsilon_s+\varepsilon'_c}d \tag{8}$$

where,

ε_s: Strain of the tensional reinforcement

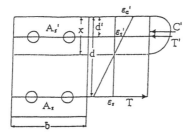

Fig. 1 Member section under flexure

Substituting this for Eq.(7)

$$S_c(\varepsilon'_c) = (p\,\sigma_s - p'\,\sigma'_s)(\varepsilon_s + \varepsilon'_c) \tag{9}$$

Differentiating both sides for ε'_c results in,

$$\frac{dS_c(\varepsilon'_c)}{d\varepsilon'_c} = \sigma'_c = \left(p\frac{d\sigma_s}{d\varepsilon'_c} - p'\frac{d\sigma'_s}{d\varepsilon'_c}\right)(\varepsilon_s + \varepsilon'_c) + (p\,\sigma_s - p'\,\sigma'_s)\left(\frac{d\varepsilon_s}{d\varepsilon'_c} + 1\right) \tag{10}$$

Here,

$$p\frac{d\sigma_s}{d\varepsilon'_c} - p'\frac{d\sigma'_s}{d\varepsilon'_c} = p\frac{d\sigma_s}{d\varepsilon_s}\cdot\frac{d\varepsilon_s}{d\varepsilon'_c} - p'\frac{d\sigma'_s}{d\varepsilon'_s}\cdot\frac{d\varepsilon'_s}{d\varepsilon'_c} \tag{11}$$

$$\varepsilon'_s = \left(1 - \frac{d'}{d}\right)\varepsilon'_c - \frac{d'}{d}\varepsilon_s \tag{12}$$

$$\frac{d\varepsilon'_s}{d\varepsilon'_c} = \left(1 - \frac{d'}{d}\right) - \frac{d'}{d}\cdot\frac{d\varepsilon_s}{d\varepsilon'_c} \tag{13}$$

where,

ε'_s: Strain of the compressional reinforcement, d': Distance to the fiber under most extreme compression from the centroid of the compressional reinforcement

Consequently, Eq.(10) can be written as follows.

$$\sigma'_c = \left\{p\frac{d\sigma_s}{d\varepsilon_s}\cdot\frac{d\varepsilon_s}{d\varepsilon'_c} - p'\frac{d\sigma'_s}{d\varepsilon'_s}\left(1 - \frac{d'}{d} - \frac{d'}{d}\cdot\frac{d\varepsilon_s}{d\varepsilon'_c}\right)\right\}(\varepsilon_s + \varepsilon'_c)$$
$$+ (p\,\sigma_s - p'\,\sigma'_s)\left(\frac{d\varepsilon_s}{d\varepsilon'_c} + 1\right) \tag{14}$$

Consequently,

$$\frac{d\varepsilon_s}{d\varepsilon'_c} = \frac{\sigma'_c - p\,\sigma_s + p'\,\sigma'_s + (\varepsilon_s + \varepsilon'_c)p'\left(1 - \frac{d'}{d}\right)\frac{d\sigma'_s}{d\varepsilon'_s}}{p\,\sigma_s - p'\,\sigma'_s + (\varepsilon_s + \varepsilon'_c)\left(p\dfrac{d\sigma_s}{d\varepsilon_s} + p'\dfrac{d\sigma'_s}{d\varepsilon'_s}\right)} \tag{15}$$

And the following formula is obtained from the ultimate limit state condition $d\varepsilon_s/d\varepsilon'_c = 0$.

$$\sigma'_{cu} = p\,\sigma_{su} - p'\,\sigma'_{su} - (\varepsilon_{su} + \varepsilon'_{cu})p'\left(1 - \frac{d'}{d}\right)\frac{d\sigma'_s}{d\varepsilon'_s} \tag{16}$$

where,

σ_{su}, σ'_{su}: Stress of the tensional and compressional reinforcement at the ultimate limit state.

By integrating formula (16) with the $\sigma - \varepsilon$ curve of the concrete, the ultimate strain of the concrete ε'_{cu} is found.

In the case of alternating cyclic loading , the ultimate limit state is caused by buckling of the compressional reinforcement (see reference [4]). It is therefore possible to assume that the compressional reinforcement fails to bear a portion of the compressive force. Consequently, the ultimate stress of the concrete σ'_{cu} is represented as follows.

$$\sigma'_{cu} = p f_{sy} \tag{17}$$

Because the ultimate strain of the concrete ε'_{cu} and $S_c(\varepsilon'_{cu})$, which is the area enclosed by the $\sigma - \varepsilon$ curve of the concrete up to the ultimate limit state, have already been found above, it is possible to find the ultimate strain of the tensional reinforcement ε_{su} from Eq.(9) as follows.

$$\varepsilon_{su} = \frac{S_c(\varepsilon'_{cu})}{p f_{sy}} - \varepsilon'_{cu} \tag{18}$$

3.1.5 Evaluating the maximum absorbed energy of the section $W_{c,max}$

The area $S_s(\varepsilon_{su})$ up to the ultimate limit state (ε_{su}) of the $\sigma - \varepsilon$ curve of the reinforcement is found as follows.

$$S_s(\varepsilon_{su}) = \frac{1}{2} f_{sy} \varepsilon_{sy} + f_{sy}(\varepsilon_{su} - \varepsilon_{sy}) \tag{19}$$

Consequently, from Eq.(2), the maximum absorbed energy of the section $W_{c,max}$ is obtained as follows.

$$W_{c,max} = (A_s + A'_s) \left\{ \frac{1}{2} f_{sy} \varepsilon_{sy} + f_{sy}(\varepsilon_{su} - \varepsilon_{sy}) \right\} \tag{20}$$

3.2 Maximum absorbed energy of the reinforced concrete member, W_{max}

Ohno et al. [5] experimentally studied the energy absorption properties of column members with a shear span of 160cm and an effective depth of 35cm. Their results revealed that approximately 90% of the energy was absorbed within 30cm of the end, which is considered the plastic region. Thus, to find the maximum absorbed energy of a member with dominant flexural behavior, considering only the plastic region causes no practical problems.

The solid line in Figure 2 represents the distribution of curvature ϕ at the ultimate limit state for columns or cantilever beams. The distribution of the strain ε_s of the longitudinal reinforcement is almost identical to the distribution of the curvature ϕ, and if the amount of energy used in producing this strain is evaluated, it is the maximum absorbed energy of the member. So if, in order to simplify the computation, the distribution of the strain ε_s of the axial reinforcement is assumed to be the dotted line in Figure 2, the maximum absorbed energy of the member W_{max} can

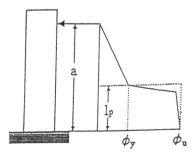

Fig. 2 Distribution of curvature ϕ at the ultimate limit state

be expressed according to the maximum absorbed energy of the section $W_{c,max}$ by the following formula:

$$W_{max} \approx l_p W_{c,max} \tag{21}$$

where,

l_p: Length of the plastic region (see Figure 2).

When the shear span ratio is large, reference [4] indicates that it is possible to estimate the length of plastic region l_p as almost identical to the effective depth (d) of the section. So if in practice, $l_p = d$, the maximum absorbed energy of the member W_{max} is as follows.

$$W_{max} = d(A_s + A'_s)\left\{\frac{1}{2}f_{sy}\,\varepsilon_{sy} + f_{sy}(\varepsilon_{su} - \varepsilon_{sy})\right\} \tag{22}$$

4 Evaluation of accumulated absorbed energy W_{acc}

It has been pointed out that damage to reinforced concrete members and structures is significantly influenced by load hysteresis [6]. Thus, before evaluating the accumulated absorbed energy, it is necessary to categorize the member by determining whether a displacement has exceeded the past maximum displacement or not.

4.1 Virgin loading point

When a displacement exceeds the past maximum displacement, points where the hysteresis loop exceeds the hysteresis loop at the time of past maximum displacement are considered virgin loading points, and their area (the inclined portion in Figure 3) is assumed to represent the damage sustained by the member. The symbols in Figure 3 are defined as follows.

δ_1: Past maximum displacement, δ_2: Present displacement,

P_y: Yield load, δ_y: Yield displacement

4.2 Cyclic loading point

If the present displacement does not exceed the past maximum displacement, the area enclosed by the hysteresis loop cannot be thought of as the damage sustained by the member. The area of the hysteresis loop that increases the damage to the member (the amount of energy) is defined as the plastic work W_p(see Figure 4), and the percentage of the plastic work Wp to the entire area of the loop W_h is assumed to be the plastic work ratio R_w.

$$R_w = \frac{W_p}{W_h} \tag{23}$$

where,

W_p: Amount of plastic work, W_h: Total work (area of one cycle of the hysteresis loop)

Uomoto et al. [7] defined the amount of plastic work W_p for displacement $i\delta_y$ (δ_y: yield displacement; $i = (\delta_i/\delta_y)$: displacement stage, however, $\delta_i > \delta_y$) as shown in Figure 4, assuming that it is the part of the hysteresis loop where the displacement

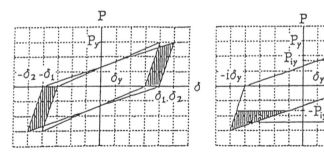

Fig. 3 Virgin loading points   Fig. 4 Plastic work

is δ_y or greater and the load is P_{iy} or greater (P_{iy}: load when the displacement is δ_y). According to this definition, the plastic work ratio R_w is determined as follows.

$$W_h = 2(i-1)P_y\,\delta_y \tag{24}$$

$$W_p = 2\frac{(i-1)^3}{(2i-1)^2}P_y\,\delta_y \tag{25}$$

Then

$$R_w = \left(\frac{i-1}{2i-1}\right)^2 \tag{26}$$

In order to find the plastic work ratio R_w, it must be hypothesized that the ultimate state is reached during the loading cycle following the virgin loading cycle (in other words, the second cycle). When monotonic loading is performed on typical reinforced concrete column members, the ductility factor at the flexural ultimate limit state is as much as 10 or more, so using the convergent value from Eq.(26) as the value of the plastic work ratio Rw presents no practical problems .

The convergent value of R_w is then found, as follows.

$$\lim_{i\to\infty}R_w = \lim_{i\to\infty}\left(\frac{i-1}{2i-1}\right)^2 = \left(\frac{1}{2}\right)^2 = 0.25 \tag{27}$$

4.3 Bauschinger effect

It is known that when reinforcement is subjected to cyclic loading, the post–yielding characteristics of its skeleton curve are different from those of the $\sigma - \varepsilon$ curve under monotonic loading. Thus the influence of the Bauschinger Effect is found experimentally from monotonic and cyclic loading tests on the same reinforced concrete specimen.

If k is substituted for the rate of increase in the energy absorbed up to failure as a result of repetitive loading,

$$W_s = \mu\cdot P_y\cdot\delta_y \tag{28}$$

$$W_c = k\left\{\mu\cdot P_y\cdot\delta_y + 0.25\sum_i(i-1)(n_i-1)P_y\cdot\delta_y\right\} \tag{29}$$

where,

W_s: Hysteresis energy until failure under monotonic loading, W_c: Hysteresis

Table 1   Experimental Results from Reference [8]

Specimen	Loading Method	Ductility Factor
2B	Monotonic Loading	7.0
2B	Cyclic Loading	3.4
8B	Monotonic Loading	7.4
8B	Cyclic Loading	3.6

energy until failure under cyclic loading, k: Rate of increase in hysteresis energy under cyclic loading, μ: Ductility factor, n_i: Number of cycles at displacement stage i

Consequently, k is found as,

$$k = \frac{W_s}{W_c} \tag{30}$$

If this is applied to the experimental results in reference [8] (See Table 1), $k=1.45$ for specimen 2B, and $k=1.40$ for specimen 8B. In principle, the value of k ought to be influenced by the load hysteresis, but there are insufficient experimental results for a proper quantification and the value of k is almost identical for the two specimens with different specifications. Therefore, here it is assumed that a uniform value of $k=1.4$ applies.

4.4 Evaluating the cumulative absorbed energy
The cumulative absorbed energy W_{acc} is the sum of the energy absorbed by the virgin loading points W_u and the energy absorbed by cyclic loading points W_l.

$$W_{acc} = W_u + W_l \tag{31}$$
$$W_u = 2kP_y(\delta_u - \delta_y) \tag{32}$$
$$W_l = 2\sum_i 0.25 n_i P_y(\delta_i - \delta_y) \tag{33}$$

where, δ_u: Ultimate displacement

5   Verification of the damage index evaluation method and its application to seismic design

5.1 Comparison with experimental results
The experimental data consisted of cases where flexural failure followed flexural yielding among the data given by Ishibashi et al. [9], Ohta et al. [10], Arakawa et al. [11], and Ohno et al. [5]. The range of specifications of the 25 specimens selected was as follows. Shear span ratio a/d: 2.5 to 4.0; longitudinal tension reinforcement ratio p: 0.36 to 1.45 (%); shear reinforcement ratio p_w: 0.08 to 0.69 (%); and axial compressive stress σ_o: 0 to 50 (kgf/cm^2). The calculated value of maximum absorbed energy is found using Eq.(22) from the specifications of these specimens, and ε_{su} in Eq.(22) is found from Eq.(18). In performing this calculation, however, both the elastic region and the energy absorbed by the concrete are ignored, which means that the results are somewhat smaller than the actual values. According to the results of

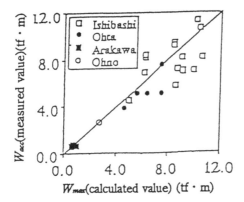

Fig. 5 Relationship between W_{max} and W_{acc}

research by Ohno et al. [5], the energy absorption ratio is 95% in the plastic region and 5% in the elastic region. An analysis by Koyanagi et al. [1] indicates that the energy absorption by the axial reinforcement is around 90% (when $p=0.5\%$, it is 93.2%; when $p=1.0\%$, it is 86.3%) and the absorption by concrete and other parts is 10%.

The experimental value for the cumulative absorbed energy at the ultimate limit state W_{acc} is found from Eqs. (31) to (33).

Figure 5, which shows the relationship between the maximum absorbed energy W_{max} and the cumulative absorbed energy W_{acc}, shows that they coincide closely. The ratio of W_{max} to W_{acc} averages 0.94, the coefficient of variation is 20.0%, and the coefficient of correlation is 0.915.

5.2 Example of analysis based on actual seismic waveforms

A reinforced concrete column with sectional dimensions 1.0×1.0 (m), a height of 4.3(m), an longitudinal tensile reinforcement ratio of 0.60 (%), and a shear reinforcement ratio of 0.25 (%) was analyzed. Two input seismic waves were used: an N−S component measured in the basement of the Japan Railway Company's Sendai Control Center Building (Miyagi−ken Oki Earthquake 1978) and El Centro (Imperial Valley Earthquake, 1940). The duration of motion was 30 seconds in all cases, the maximum acceleration ranged from 400 gal to 700 gal, and the recorded earthquake waves were magnified and contracted.

From the specifications of the specimen, the maximum absorbed energy W_{max} was computed as 2.05×10^{6} (kgf·cm). The cumulative absorbed energy, on the other hand, was found from a single degree of freedom elasto−plastic dynamic response analysis. The results are shown in Table 2.

This table shows that the damage index value D did not reach 1.0. In other words, this column did not reach the ultimate limit state, even though the maximum acceleration was increased to 700 gal. The damage index D rose sharply at an acceleration of 500 gal or more, where plastic deformation took place. This result indicates that the structure could collapse if it is subjected to several earthquakes of this magnitude.

Table 2   Calculated Values of Damage Index D from Actual Seismic Waves

Maximum Acceleration	Miyagi−ken Oki		El−Centro	
(gal)	W_{acc}	D	W_{acc}	D
400	0.0107	0.005	0.0306	0.015
500	0.149	0.073	0.196	0.096
600	0.674	0.329	0.854	0.416
700	1.19	0.581	1.36	0.664

5.3 Probabilistic evaluation of the damage index

5.3.1 Damge index and failure probability

The cumulative absorbed energy input by the earthquake W_{acc} has a fixed value, and the maximum consumed energy W_{max} is hypothesized in such a way that by assuming that the section area of reinforcement A_s and the effective depth of the section d have fixed values, and that $S_s(\varepsilon_{su})$ is a random variable of the logarithmic normal distribution (because it is not negative) with a coefficient of variation of 10%, the probability distribution of the maximum absorbed energy W_{max} is the same as the probability distribution of $S_s(\varepsilon_{su})$ (See Eq.(22)). Then, the cumulative absorbed energy W_{acc} corresponding to $p_f = 10^{-6}$, which is the failure probability p_f value typically allowed for important structures, is found.

The failure probability p_f can be represented as follows using the damage index D.

$$P_f = P(D \geq 1.0) = P(W_{acc}/W_{max} \geq 1.0) = P(W_{acc} \geq W_{max}) \tag{34}$$

If W_f is substituted for the cumulative absorbed energy W_{acc} corresponding to the failure probability $P_f = 10^{-6}$.

$$P(W_f \geq W_{max}) = 10^{-6} \tag{35}$$

If W_{max} is converted to the standard normal distribution,

$$10^{-6} \fallingdotseq \Phi(-4.75) \tag{36}$$

Therefore,

$$\Phi\left(\frac{\ln W_f - (\ln E(W_{max}) - 0.00488)}{0.0988} \right) = \Phi(-4.75) \tag{37}$$

where,

$\Phi(x)$: Standard normal distribution function

Consequently,

$$\ln W_f - \ln E(W_{max}) = -0.469 \tag{38}$$

$$\frac{W_f}{E(W_{max})} = e^{-0.469} \tag{39}$$

$$W_f = 0.63 \cdot E(W_{max}) \tag{40}$$

Because the average value $E(W_{max})$ of the maximum absorbed energy W_{max} is the calculated value obtained from the specifications of the cross section, the damage index at this time is calculated as follows.

Table 3 Failure Probability P_f – Damage Index D Relationship

Failure Probability P_f	10^{-3}	10^{-4}	10^{-5}	10^{-6}
Damage Index D	0.74	0.69	0.66	0.63

$$D=\frac{W_f}{W_{max}}=0.63 \tag{41}$$

Table 3 presents a summary of damage index values D corresponding to various failure probabilities found by the hypotheses and procedures described above. The failure probability P_f of a structure corresponding to a damage index value D is found using this table. This allows designers to determine the damage index D allowed for a structure and evaluate the reliability of the structure.

5.3.2 Application to seismic design
Figure 6 is a flow chart for a seismic design method based on the damage index. The first step is to determine the normal variability of the maximum absorbed energy W_{max} and the cumulative absorbed energy W_{acc}. Based on the results, the value of the damage index D is then found from the failure probability P_f of the structure. Next, the cumulative absorbed energy W_{acc} for the construction site is calculated. When doing so, it is possible to adjust this value to take into account the ground and other conditions. When W_{acc} and D have been found, W_{max} can be determined, and the design values are then based on this figure. It is possible to apply the damage index concept to seismic design by using the above procedure.

6 Conclusions

This study proposes a method of evaluating a damage index for reinforced concrete members that suffer flexural yielding followed by flexural failure. The following are the main conclusions reached in the study:

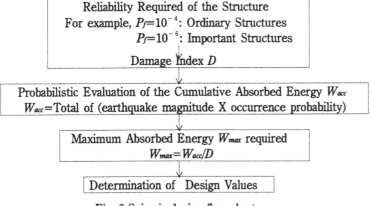

Fig. 6 Seismic design flow chart

(1) The study defined a damage index for a reinforced concrete structure by treating the amount of energy as an index and showed how to evaluate this damage index.

(2) Through the analysis of real seismic waveforms, the damage index was calculated, demonstrating that this index could be used as a measure of the reliability of a reinforced concrete structure.

(3) The study described a probabilistic approach to the evaluation of the damage index and set out a reinforced concrete structure seismic design method using the damage index.

7 References

1.   Koyanagi, W., Rokugo,K., and Iwase, H.(1984) Effects of Material Properties on Flexural Failure Process and Flexural Toughness of Reinforced Concrete Beams. *Proc. of JSCE*, No.348／Ⅴ−1, pp.153−162, (in Japanese)

2.   Muguruma,H., Watanabe,F., Katuta,S., and Tanaka,H.(1980) Modeling of Stress−Strain Relationship of Confined Concrete, *CAJ, Proc. of Cement Technology*, Vol.34, pp.429−432, (in Japanese)

3.   Suzuki,K., Nakatsuka,T., and Inoue,K.(1988) Ultimate Limit Index Points and the Characteristics of Partially Prestressed Concrete Beam Sections, *JCI Colloquium on Ductility of Concrete Structures and its Evaluation*, pp.Ⅱ. 193−204, (in Japanese)

4.   Shima,H., Itoh,K., Kitanishi,T., and Mizuguchi,H.(1990) Effect of Hoop Arrangement on Bar Buckling and Ductility of Reinforced Concrete, *Proc. of JCI Symposium on Design of Reinforcement and Ductility of Concrete Structures*, pp.33−40,

5.   Ohno,T., and Nishioka,T.(1984) An Experimental Study on Energy Absorption Capacity of Columns in Reinforced Concrete Structures, *Proc. of JSCE, Struct.Eng./Earthquake Eng.*, Vol.1, No.2, pp.23−33,

6.   JCI,(1988) *Committee Report on Ductility of Concrete Structures and Its Evaluation*, (in Japanese)

7.   Uomoto,T., Yajima,T., and Hongo,K.(1992) Method for Calculating the Total Dissipated Energy of Simple Reinforced Concrete Beams Failing by Flexure under Cyclic Deformations, *Proc. of the Japan Concrete Institute*, Vol.14, No.2, pp.919−924, (in Japanese)

8.   Azuma,Y., Okubo,D., and Otsuka,M.(1975) Study on Prevention from Failure of RC Short Columns, *Proc. of Annual Conference of JAI*, pp.1113−1114,

9.   Ishibashi,T., and Yoshino,S.(1988) Study on Deformation Capacity of Reinforced Concrete Bridge Piers under Earthquake, *Proc. of JSCE*, No.390 ／Ⅴ−8, pp.57−66, (in Japanese)

10.   Ohta,M.(1980) A Study on Earthquake Resistant Design for Reinforced Concrete Bridge Piers of Single−Column Type, *Report of Public Works Research Institute*, No.153, (in Japanese)

11.   Arakawa,T., Arakawa,Y., Fujita,Y., and Mizoguti,M.(1981) Revaluation of Deformation of RC Columns under Cyclic Loading, *Proc. of the Japan Concrete Institute*, Vol.3, pp.449−452, (in Japanese)

20 ENVIRONMENTAL ISSUES FROM THE CONSTRUCTION INDUSTRY

Y. NOJIRI
Kajima Technical Research Institute, Tokyo, Japan

Abstract

This paper provides an outline of the relationship between construction work and the global environment. Also presented here are the state-of-the -art measures being undertaken by construction-related bodies in Japan for the conservation of the global environment.

Keyword: Global environment, construction waste, recycling, formwork, tropical rain forest, carbon dioxide.

1 Introduction

The present and future survival and progress of mankind are placed in jeopardy by the deterioration of the global environment, in the form of global green house effect, destruction of the ozone layer, acid rain and destruction of the tropical rain forests. At the root of these problems lie the pursuit of material comforts in the developed countries and the population increase in the developing countries, which have resulted in the consumption of natural resources and production of waste in excess of the capacity of the environment to sustain them. If the crisis we now face is to be averted, these problems must be addressed from a variety of viewpoints. In Japan too, at the levels of the national and local governments, private enterprises and individual citizens, all are endeavoring in accordance with his or her status and condition, in activities aimed at coexistence of environmental conservation and economic development.

The reduction in the consumption of natural resources and generation of waste is an urgent task also for those of us engaged in construction, and on the understanding that the solution of these problems will also lead to economy in construction industry, active research is under way for the development of technologies aimed at the reduced con-

Integrated Design and Environmental Issues in Concrete Technology. Edited by K. Sakai. Published in 1996 by E & FN Spon, 2–6 Boundary Row, London, SE1 8HN, UK. ISBN 0 419 22180 8.

sumption of resources and energy, and reduction and recycling of the by-products of construction.

Among the various environmental problems, this paper deals mainly with the problems of construction by-products, the protection of tropical rain forests and the green house effect, providing the current situation in Japan, together with an outline of the efforts being made by the construction-related bodies including the construction industry for these environmental issues and the results of the technological development work implemented so far.

2   Construction and environment

The relations between the global environmental issues and the measures being undertaken in the field of construction are summarized in Table 1[1]

Table 1 Global environmental issues and construction [1]

Legend ● : very strong correlation   O : strong correlation   ○ : correlation observed

Global environmental issues (AIJ) \ Measures taken by construction industry	Prolongation of service life of structures	Provision of greenery in urban areas and around buildings	Review of uses of concrete	Review of uses of steel	Review of uses of wood	Energy-saving buildings	Alternative sources of energy	Review of emission of CFCs	Review of use of plywood made from tropical wood	Rationalized disposal of construction by-products	Rationalized transportation of construction materials	Reduced use of paper in construction activities
Global warming	●	○	●	●	●	●	●	○				○
Destruction of ozone layer							○	●	◎			
Atmospheric pollution			○	○		◎	◎				◎	
Acid rain			◎	◎		◎	◎				◎	
Heat islands		◎				◎					○	
Marine pollution											○	
Diffusion of toxic matter											○	
Desertification		○			○					○		
Reduction of forests	◎				○					◎		○
Reduction of wildlife species					○				●			
Increased exploitation of resources and production of waste	●		◎		◎					◎	○	

3   Reduction of construction by-products

It is suggested that the resources consumed and waste produced be reduced by direct measures, such as the improvement of the disposal methods for and recycling of construction by-products and the reduction of the use of concrete, wood and other construction materials. Furthermore, in addition to the those direct measures, indirect measures for reduction of waste materials in the form, for example, of the extension of the service lives of structures themselves through the development of materials, design and construction methods aimed at greater durability are of importance.

3.1 Action by the ministry of construction
The fundamental concepts relating to reducing construction by-products are based on the following three principles [2] [3]
(a) Reduction of waste materials through improvement of construction procedures
(b) Promotion of reuse as construction materials
(c) Enforcement of appropriate disposal in landfills
Construction by-products can be defined here as "materials created as secondary products in construction work" and encompasses a wide range of materials, including such materials as sludge, waste paper, metal shavings and glass in addition to the earth removed from the construction sites (surplus soil), concrete masses, asphalt concrete masses and wood (waste wood and wood shavings).

Construction by-products are classified into recycled materials and waste materials, to each of them, different laws and regulations apply. The relationship between construction by-products, recycled materials and waste materials is shown in Fig. 1. Materials which cannot be used as raw materials do not fall under the category of recycled materials. For recycled materials, the law concerning the promotion on the use of recycled materials (Recycling Law) was drafted in accordance with Principle b) above by the Ministry of Construction, Ministry of International Trade and Industry and five other ministries and agencies and was enforced in October 1993 [4]. Under this law, the construction industry is named as one of the designated industries responsible for promoting the use of recycled materials. The listed materials to be recycled are as follows.

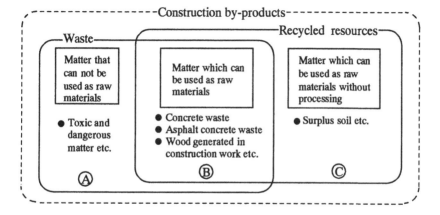

Fig. 1 Construction by-products, recycled materials and waste materials[2]

(a) Surplus soil (earth)
(b) Concrete waste
(c) Asphalt concrete waste
(d) Waste wood generated in construction work

The Recycling Law stipulates separately the provisions concerning the use of re-cycled resources as construction materials, such as recycled crusher runs and recycled hot asphalt mixtures obtained at the construction sites and surplus soil brought in from other sites. Also the provisions aim at facilitating the use of recycled resources at other construction sites of construction by-products such as earth, concrete and asphalt concrete masses produced at construction sites. This separate treatment of the two reminds one that there are an inflow and outflow of recycled materials at each construction site. Waste materials are provided for under the Waste Disposal and Public Cleansing Law (Waste Disposal Law).

3.2 Actions by construction industry [5]

According to a survey conducted by the Japan Civil Engineering Contractors' Associa-tion, energy consumed directly by the construction industry (i.e. energy consumed in the actual construction work) accounts for 1.6% of the total energy consumption in Japan, while the indirect consumption (by transportation, steel, ceramic, soil and stone, chemi-cal and other industries) accounts for 18.6%. The two together add up to a total of 20.4%. In other words, the direct effect, in terms of energy consumption, of the construction industry on the environment is small. The considerations, therefore, on the methods of reducing energy consumption in the field of construction should not be limited to the stage of actual construction, but should cover the whole life cycle of structures from the stage of raw materials to demolition and disposal.

Under these circumstances, starting in 1989, the Japan Federation of Construction Contractors and other bodies in the construction industry have been establishing com-mittees (e.g. the Special Committee on Global Environmental Issues) for the purpose of conducting studies on the restoration, conservation and creation of natural environments. Through these committees, the role of the construction industry has been reviewed re-garding the promotion of environmental policies and the activities the construction in-dustry should engage in future. It is to be added that by nature, the construction industry is categorized as "environmental industry".

As a part of this work, the Japan Federation of Construction Contractors drew up the guidelines for preparing the part of each member company regarding their environmen-tal conservation programs. These guidelines specify the activities that should be included in the environmental conservation programs as follows.

1. Quantitative assessment of the reduction of environmental load and effective utilization of natural resources
2. Development of technology and products for achieving 1) above and technology transfer for this purpose
3. Education of those engaged in construction and provision of information
4. Support for environmental conservation activities in regional communities
5. Environmental conservation activities in the relevant countries and societies in overseas projects
6. Establishment of appropriate organization within each companies

In addition to the above guidelines concerning basic policies, the Building Contractors' Society has proposed, as an objective target, the reduction of the consumption of plywood forms for concreting made from tropical wood by 35% from the present level within the next five years.

To sum up, the construction contractors are engaged at present in the work of drawing up their action programs and making proposals for the actual implementation of environmental conservation measures in accordance with the concepts for environmental protection presented by the national government and the Ministry of Construction in particular. Also, the policies for drawing up environmental conservation plans issued by various construction-related bodies are being done.

3.3 Construction waste

As shown in Fig. 2 [2] the total volume of construction waste generated in 1990 was estimated at 76 million tons. Construction waste has seen a rapid increase in recent years, with an increase of 89% between 1980 and 1985 and 33% between 1985 and 1990. These are particular large increases when compared with the increase of only 7% in all industrial waste between 1985 and 1990. With the shortage of final disposal yards, it is becoming difficult today to dispose of construction waste.

Since construction waste is generally very stable in their properties with the exception of a very limited number of items such as dispersible asbestos, it is thought that one should attempt to recycle as much of it as possible. Only 35% of construction waste, however, is being recycled at present and, even if taking the reduction in its volume by dehydration and incineration into account, 58% of construction waste is being disposed of as it is at the final disposal yards.

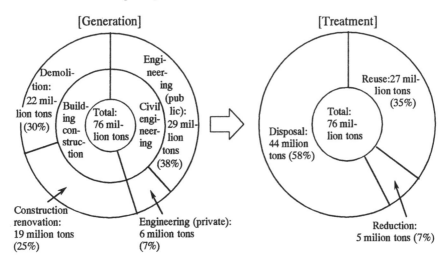

Fig. 2 Generation and disposal of construction waste[2]

The volumes of construction waste are shown according to items in Fig. 3 [2]. Concrete masses account for the largest proportion (25 million tons, 34%), followed by asphalt concrete masses (18 million tons, 23%) and construction sludge (14 million, 19%). Concrete masses are either reused as recycled concrete aggregates or disposed as industrial waste in landfills. Concrete masses generated in demolition of underground structures, since they contain mixtures of soil and therefore require screening, are generally disposed as soil containing concrete masses. Asphalt concrete masses are recycled after treatment such as heating and mixing, and the residual portions are disposed in landfills. Mixed waste is screened for recovery of usable materials and the residual materials are compressed and disposed in landfills, while construction sludge is disposed in landfills after dehydration. While the recycling rate exceeds 50% for asphalt concrete masses and concrete masses, it remains low for surplus wood (31%), mixed waste (14%) and sludge (8%), necessitating an improvement of the recycling rate in future.

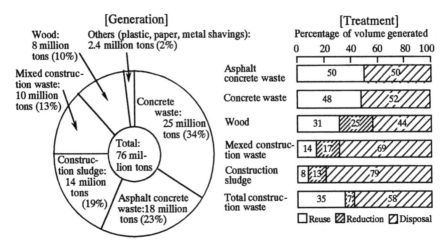

Fig. 3 Generation and disposal of construction waste according to items [2]

3.4 Use of concrete waste as recycled aggregate
The qualities of recycled aggregate are given in Tables 2 [6] and 3 [6] with reference to the types of the original concrete and levels of crushing treatment. It may be seen that materials obtained from high-strength concrete do not necessarily provide high-quality

Table 2 Properties of recycled aggregate [6]

Treatment level	Strength of original concrete	Specific gravity	Absorption (%)	Attached mortar (%)
Primary crushing	High strength	2.48	4.02	41.7
	Low strength	2.38	5.18	29.4
Intermediate	High strength	2.57	2.48	26.5
	Low strength	2.55	2.56	20.6
High	High strength	2.62	1.55	10.8
	Low strength	2.66	1.06	2.5

High strength : approx. 400kgf/cm^2
Low strength : approx. 200kgf/cm^2
Primary crushing : crushing of original concrete only

recycled aggregate. In fact, the recycled aggregate of higher quality results from low-strength concrete in the case of fine aggregate. This may be attributed to the fact that the lower the strength of the original concrete, the easier it is to remove the mortar and paste. In this way, the qualities of recycled aggregate are influenced by the qualities and quantities of the mortar attached to the aggregate materials. A useful index for the quality of the recycled aggregate may therefore be the absorption of the original concrete, which is closely related to the quality of the mortar.

Table 3 Properties of recycled fine aggregate [6]

Treatment level	Strength of original concrete	Specific gravity	Absorption (%)	Attached morter (%)
Intermediate	High strength	2.37	7.84	29.0
	Low strength	2.41	6.58	15.3
High	High strength	2.45	5.86	23.3
	Low strength	2.52	3.76	9.3

Experimental research concerning the effective use of recycled aggregate was conducted by the "Committee for Development of Technologies for Utilization of Waste Materials in Construction Work" (Japan Institute of Construction Engineering) between 1981 and 1985, leading to the publication of the "Guidelines for Design and Execution of Concrete Using Recycled Aggregate (Draft)" [7] by the Civil Engineering Subcommittee, a subgroup of the committee. On the assumption that the conditions when making partial use of recycled materials will be on the safe side in comparison with when using 100% recycled aggregate, figures are given in these guidelines for cases where all the fine and coarse aggregates consist solely of recycled materials. The qualities of the recycled aggregate and the types of concrete produced with recycled aggregate as given in the guidelines are shown in Tables 4 and 5 below.

Table 4 Types of recycled coarse aggregate concrete [7]

Type	Recycled coarse aggregate			Recycled fine aggregate	
Item	Type 1	Type 2	Type 3	Type 1	Type 2
Absorption (%)	≤ 3	≤ 3	≤ 5	≤ 7	≤ 5	≤ 10
Stability	≤ 12	≤ 30 (≤ 40)*	≤ 12	–	≤ 10	

* : when freezing and thawing resistance not considered

Table 5 Qualities of recycled coarse aggregate [7]

Recycled concrete type	Design strength of recycled concrete σ_{ck} (kgf/cm^2)	Coarse aggregate used	Fine aggregate used
I	≥ 210 (reinforced concrete)	Type 1 recycled aggregate	Ordinary aggregate
II	≥ 160 (plain concrete)	Type 2 recycled aggregate	Ordinary aggregate or Type 1 recycled aggregate
III	≥ 160 (sub-slab concrete)	Type 3 recycled aggregate	Type 2 recycled aggregate

4   Protection of rain forests and formwork problems

Both forests and oceans play an important role for the conservation of the global environment from the aspects of recycling materials and energy. Trees can absorb and fix carbon dioxide which promotes so called green house effect and results in global warming; particularly, tropical rain forests serve as a vast storehouse for carbon dioxide. Tropical rain forests are also a repository of wild animals, which in turn are valuable genetic resources necessary for improving cattle breeding and developing of pharmaceutical products. In developed countries, deforestation and afforestation of coniferous trees are almost balanced with a slight increase in the area of the forests; however, regeneration of rain forests by afforestation is difficult. Due to large-scale agricultural developments, use of wood as fuel, and commercial deforestation, the area of forests is rapidly decreasing; the rate is estimated as 17,000 km^2 every year.

The percentage of deforested tropical wood used in concrete formwork in the construction industry in Japan is 14% of the total imported quantity, as shown in Fig. 4. This quantity is not excessive. However, since the wood is thrown away or disposed after the formwork. the use of wood in formwork is by itself regarded as a problem for which a solution is demanded considering the aspect of appropriate use of tropical wood.

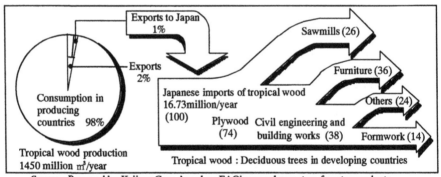

Source : Prepared by Kajima Corp. based on FAO's annual report on forestry products
for 1987 and reports of the Japan Plywood Manufacturers Association for 1987.

Fig.4   Percentages of deforested tropical wood used in formwork

4.1 Actions by governmental agencies and other associations

The Ministry of Construction prepared guidelines called "Future Trends for Rational Use of Plywood in Formwork" in August 1991 in response to the problem of reduction of tropical forest area.   The guidelines promote the use of plywood made of tropical timber and coniferous timber as a short-term measure, and the use of alternatives for wood in formwork, promotion of the development and application of alternative work methods as long-term measures. The Ministry has also started ordering model works since 1992.

Several municipal governments have prepared guidelines for reducing the use of tropical wood and have indicated specific numerical targets to be attained, such as the reduction in the percentage of tropical wood used in work contracted by the government to 30% within five years. In view of such actions, the Building Constructors Society (BCS) also investigated the status of usage of plywood construction materials. In January 1991, based on the results of the investigations, the society summarized targets and measures through planned reduction in the consumption of tropical wood used in formwork, and is in the process of adopting comprehensive countermeasures.

4.2 Plans for reducing consumption of tropical wood

Plans for reducing consumption of tropical wood may be broadly classified into: (1) repeating usage of formwork, (2) changeover to formwork made of alternative materials, and (3) changeover to construction methods that do not require formwork, as shown in Table 6. From these categories, plywood can be considered for short-term usage, and precast formwork (stay-in-place formwork) made of alternative materials may be considered for long-term usage. Recently, in addition to these two measures, plastic panels to be recycled have been developed for formworks. This is produced by extrusion molding using polypropylene resin as the raw material and can be totally recycled.

Table 6 Methods for reducing consumption of tropical wood

Classification of plan	Details
Repeated use of formwork	Use of plywood with high durability,such as coated plywood
	Rational design of structures. (consider minimizing the cutting and processing work in design for re-use)
Changeover to formwork made of alternative materials	Use of plywood made of coniferous wood
	Use of formwork made of other altenative materials
	Use of precast formwork (deck plate, thin PC sheets, etc.)
Changeover to construction methods that do not require formwork	Promote prefabrication of concrete structual members
	Cangeover to methods that do not use formwork

4.3 Plywood

Plywood is formed by cutting wood into thin plies (single sheets), assembling them, and bonding them together using adhesive to form a single panel. In a panel made of coniferous wood, there were defects such as transfer of growth rings, knotholes, and resin on the surface; moreover, some kind of coniferous wood affects the hardening of concrete surface. On the other hand, plywood combined with tropical timber and conifer ply, as shown in Fig.5, can compensate for these defects and this results in reduction of use of tropical timber by 70%. Tests using formwork made of the plywood gave rather irregular surfaces, but the strength obtained was adequate for use in formwork; this material could be used satisfactorily in most of the members without problems.

JAS (Japan Agricultural Standard) approval for the plywood for use in formwork has already been obtained. Since the cost is almost the same as tropical timber, this material is likely to be used extensively as a substitute to panels made of tropical timber.

4.4 Precast (stay-in-place) formwork

Research and development on precast concrete formwork, which is not removed after placing concrete and becomes a part of the surface, is being carried out vigorously in

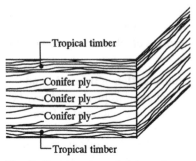

Fig. 5 Construction of plywood

various construction companies in recent years. Several products incorporating various contrivances have already appeared in the market. New techniques are attracting considerable attention today because the generation of waste can be minimized, work period can be shortened and labor savings can be attained. These techniques are of two types:

(a) Thin-walled formwork used as stand-alone formwork

(b) Thick-walled formwork used as a part of the structure, containing
      main reinforcements within the formwork

The technique in (b) above is rational, and the reinforcement assembly work can be carried out in the plant; therefore, labor savings at construction sites can be attained. However, there are disadvantages, such as the problem of joints of main reinforcement and formwork, and the increase in weight of the formwork. This technique is currently being developed in the Project for Resolving Various Construction Issues at the Ministry of Construction, and site tests are being carried out mainly for high bridge piers. (See Fig. 6)

The technique in (a) has the disadvantage of increased cross section area of members, However, in addition to labor savings in on-site work, durability and appearance improves; therefore, various types of formwork using cement-based material with a high compressive strength of about 100 Mpa are being developed.

Almost all the precast formwork techniques today result in increased costs compared to existing techniques, and they have not become the mainstream formwork techniques; however, it is anticipated that such techniques will be increasingly used in the long term together with the need for improved productivity at plants and sites.

Fig. 6 is showing the use of large precast concrete forms applied for the Kurami bridge project.

5   Concrete and carbon dioxide problems

In recent years, the awareness that global environmental problems affect the lives of human beings is increasing, as exemplified by the United Nations Conference on Environmental Development (UNCED: Global Summit) held at Rio de Janeiro in June 1992. Global warming, one such environmental problem, is being treated as a serious problem because of the increasing discharge of carbon dioxide into the atmosphere. In Japan too, the "Action Plan for Preventing Global Warming" was adopted at the Cabinet meeting in October 1990, and the target of "stabilizing carbon dioxide emission per person to 1990-levels after the year 2000" was set.

Use of large precast formwork in the Kurami bridge

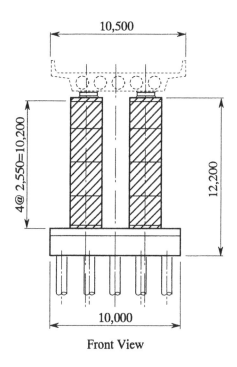

Front View

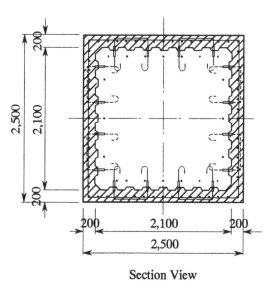

Section View

Fig.6 Pre-cast formwork for bridge under development in the project for resolving various construction issues

5.1 Present status of carbon dioxide emission in construction industry

Fig.7 [8] shows the carbon dioxide emission by industry in Japan in 1990. As seen in this figure, the carbon dioxide emission from the cement industry, an indispensable part of the construction industry accounts for 5.7% of the total emission for the whole country, including emission from raw materials. According to other statistics, the carbon dioxide emission from construction-related industries is estimated as 11% of the emission from all industries. Neither of these figures can be ignored when considering the reduction of carbon dioxide emission; various measures are being adopted from different aspects.

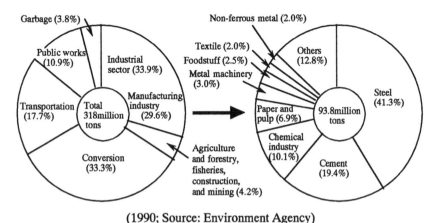

(1990; Source: Environment Agency)
Fig. 7 Direct emission of carbon dioxide by industry in Japan

5.2 Measures to reduce carbon dioxide emission

Issues on carbon dioxide associated with the construction industry are numerous; they include issues that are directly associated with the industry, such as discharge of gases from construction machinery, and indirect associated such as deforestation for construction of structures. However, the amount of carbon dioxide emissions associated with cement, and the main raw material for concrete, is generally considered to be high; in this report, only measures for reducing carbon dioxide emission associated with concrete are described.

5.2.1 Cement manufacture

Energy conservation measures for reducing carbon dioxide emission currently being considered in the cement industry include the following:
(a) Use of more efficient cement manufacturing equipments
(b) Re-use of industrial wastes and by-products during cement manufacture
(c) Promotion of use of blended cement
With regard to (a), a considerably high energy efficiency has already been attained; it is anticipated that future technological developments will not offer significant improvements.

On the other hand, the increased use of blended cement with the focus on blast-furnace slag is an excellent measure and has the maximum effect on consumption.

In recent years, many researches have been carried out on the manufacture and re-use of bulk wastes such as ash from city refuse incinerators and sewage sludge [9]. These wastes are being investigated because they contain SiO_2, Al_2O_3, Fe_2O_3, and CaO, which form the basic constituents for the manufacture of cement. By adjusting the composition of clinker, various types of cement ranging from Ordinary Portland Cement to extra rapid hardening Portland cement can be manufactured. This is called "Eco-Cement", so far in Japan. At present, the use of such cement is restricted to non-reinforced concrete because of the inclusion of chlorides in the raw material. Results of researches have just been reached indicating that this is one of the best measures for recycling resources.

5.2.2 Greening of concrete

Trees absorb carbon dioxide, are effective in purifying the air, retain rain water, and assist in recovery of the ecological system. Furthermore, trees ease psychological stress in city dwellers, and provide a soothing effect. In view of these advantages, tests are being carried out vigorously for the greening of concrete structures in cities where functionality and economy are given precedence above everything else.

Greening methods are classified into two categories: that is, (a) the method of bringing soil from another place, placing it in a hole opened in a part of the structural member, and planting trees in this soil; and (b) the method of planting trees directly in concrete. The former includes the planting of trees in a building roof or atrium in buildings; the latter includes the planting of trees on a slope of embankments. In the latter method, porous concrete is used in which continuous pores are created so that water and air can circulate freely; research on mix and experimental investigations on growing plants have been carried out until now. The realization of this method of directly planting trees in concrete is anticipated. Developments on the 'greening' of concrete have just started; aesthetic design methods, such as development of hardware techniques together with investigations on where and how to grow plants need to be carried out.

5.3 Durability of concrete and carbon dioxide

Methods for reducing carbon dioxide and its association with the concrete industry were studied. The various measures described here for reducing carbon dioxide should be implemented, but more important is the reduction of total usage volume from a long term perspective by increasing the life span of structures. In recent years, high range water reducing agents and various admixtures have been developed. By using these new materials, concrete materials with high performance, high functionality, and various new technologies for improved durability have been developed and realized. To apply these technologies to ordinary concrete structures and to build structures of excellent durability, it is necessary to devise systems for evaluating the applications of these new technologies and new materials.

6   Concluding remarks

Based on the previous and current studies, the relationship between construction work and the global environment, and a state-of-the-art on the measures being undertaken in Japan for the conservation of the global environment as a part of the endeavor of construction industry, together with some ideas for promoting the recycling of construction by-products were described on this paper. The author would like to express his appreciation for the researchers of the studies quoted.

7   References

1. Building Contractors' Society (1992) Special Committee on Global Environmental Issues, *Results of the Survey on the Environmental Load due to Construction Activities in Japan and Related Activities, and Directions of Future Activities by the Construction Industry*.
2. Economic Affairs Bureau, Ministry of Construction (1992) *Construction Industry and Measures Relating to Construction By-Products*.
3. Kurata, M. (1992) "Present State and Future Prospects for Recycling of Construction By-Products", CIVIL ENGINEERING JOURNAL (doboku-gijutsu-shiryo) 34-11, pp.32-39
4. Okudaira, H. (1992) " The Recycling Law - the Aims of the Ministry of Construction", CEMENT CONCRETE No. 548, pp.36-41
5. Environment Committee of the Japan Federation of Construction Contractors (1992) *The Global Environment and the Construction Industry*.
6. Kawano, H. (1986) " Concerning the Guidelines for Design and Execution of Concrete Using Recycled Aggregate (Draft)", CIVIL ENGINEERING JOURNAL (doboku-sekou).
7. Ministry of Construction & Japan Institute of Construction Engineering (1986) *Report of the Studies concerning the Development of Technologies for Application of Waste Materials to Construction Work.*.
8. Iizuka, S. (1994) "Cement manufucture and environment", CEMENT CONCRETE, No.563
9. Uchida S.pp.2-1－2-8 "Eco-cement," Lecture note of the working group on cement, Japan Ceramics Association

21 EFFICIENT USE OF RESOURCES IN THE FIELD OF CONCRETE TECHNOLOGY

K. ROKUGO and N. KUJIHARA
Gifu University, Gifu, Japan

Abstract
The ISO standards on the quality assurance, the environmental performance and the performance requirement for concrete structures are outlined. It is pointed out that the extending the life span of concrete structures is essential from a viewpoint of the environmental performance. The attempts and ideas for the efficient use of resources in the filed of the concrete technology are classified and several of them are introduced in order to discuss new directions of the technology.
Keywords: Concrete technology, demolished concrete, efficient use of resources, ISO standards, recycled aggregate

1 Introduction

The concrete technology has contributed to the design and construction of safe, durable and economical concrete structures. The ease in both construction and maintenance is needed for concrete structures. The importance of the quality assurance and the environmental performance in the design and construction of concrete structures is increasing.

In this paper, the ISO (International Standardization Organization) standards on the quality systems (ISO 9000 series) and the environmental management systems (ISO 14000 series), as well as the ISO performance code for structural concrete, are briefly described. Many attempts and ideas of efficient use of resources in Japan are classified and several of them are introduced in order to discuss new directions of the concrete technology.

Integrated Design and Environmental Issues in Concrete Technology. Edited by K. Sakai. Published in 1996 by E & FN Spon, 2–6 Boundary Row, London, SE1 8HN, UK. ISBN 0 419 22180 8.

2   ISO standards and performance code

The adjustment of JIS (Japanese Industrial Standard) standards to ISO standards is accelerated also in the field of the concrete technology, namely for cement, concrete, reinforcing and prestressing steel, etc.

In the ISO standard on the quality systems, the quality assurance is defined as "all those planned and systematic actions necessary to provide adequate confidence that a product or service will satisfy given requirements for quality." The first certification for the quality systems will be given to a branch of a general contractor in Japan by the end of this year (1995) by JTCCM (Japan Testing Center for Construction Materials).

The ISO standard on the environmental management systems is scheduled to be established in the coming spring (1996). This standard aims to reduce the load against the environment. In the case of concrete structures, extending the life span of structures is essential.

ISO codes for the design of concrete structures are discussed in ISO/TC71. ISO/TC71/SC2 intends to issue an operational code (level-2) such as Eurocodes. On the other hand, ISO/TC71/SC4 aims to establish a performance code (level-1), where principles and procedures with serviceability and safety levels are expressed. The performance code will be structured so that the code provisions would be attainable by the various other code groups. It consists of requirements, criteria and assessment. It would be possible that the quality assurance and the environmental performance will have a greater role in the future ISO codes for structural concrete.

3   Efficient use of resources

3.1   Classification of attempts and ideas

The attempts and ideas for the efficient use of resources in the field of concrete technology can be classified into two groups, "utilization of resources" and "restraint of consumption of resources", as shown in Fig. 1. The former contains the so-called recycling such as the use of demolished concrete, the development of unused resources, and the utilization of wastes and by-products from other fields. The latter includes the reduction of the occurrence of construction wastes, the reduction of the quantity of construction materials, the restraint of the energy consumption and the prolongation of the life span of concrete structures, where techniques for the durability design, the maintenance, the rehabilitation and the life-extension are important. After a description of the construction waste and the disaster wastes, several topics dealing with the efficient use of resources are presented and discussed.

3.2   Construction waste and disaster waste

The total amount of construction wastes from architecture and civil engineering field in Japan in 1990 was 76 million tons [1, 2] (Fig. 2). Fifty five percent was produced in the architecture field and 45% in the civil engineering field. Out of 76 million tons, 58% was not reused but disposed. The current annual amount of the concrete waste is 25 million tons. Half of them is disposed and the other half is reused [1].

Many structures were destroyed due to the Hyogoken-Nanbu earthquake on

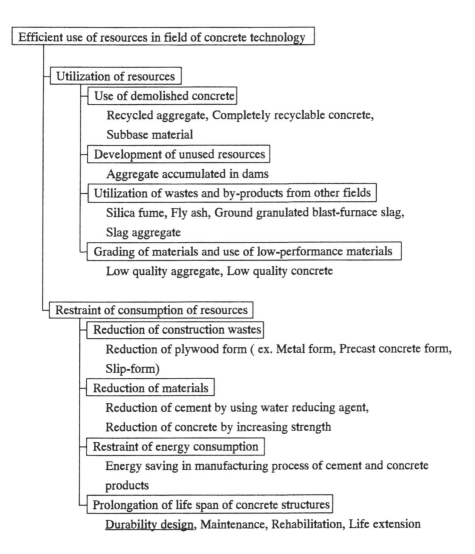

Fig. 1 Efficient use of resources in field of concrete technology

January 17 th, 1995.   According to the publication on January 26th, the amount of disaster waste from houses and buildings was about 6 million tons (5 million m³) and the amount from public facilities such as roads, railways and harbors was about 5 million tons (3 million m³).   The reuse of the disaster waste, including demolished concrete, seems to be difficult, because the huge amount of wastes appeared at once in the small area.

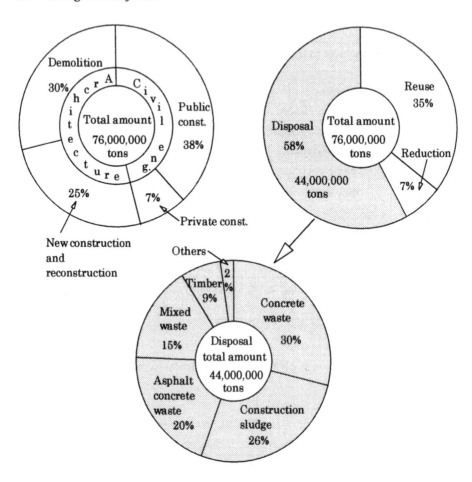

Fig. 2   Construction wastes in 1990

3.3   Reuse of demolished concrete

3.3.1   Subbase material

Most of the reused demolished concrete is used as subbase material.   Yoshikane [3] reported that the quality of the reused subbase material does not depend on the compressive strength of the source concrete and that the recycled material is as strong as the cement-stabilized soil, due to the re-hydration of unhydrated cement in the material.   Moreover, the addition of the sludge from ready mixed concrete is effective to increase the strength of the material.   There is a limit in the consumption as the subbase material.   The amount of demolished concrete increases year by year.   The opening gap will be a problem.

Table. 1   Tentative specification for quality of recycled aggregate

	Recycled coarse aggregate			Recycled fine aggregate	
Classification	1st	2nd	3rd	1st	2nd
Absorption(%)	≤ 3	≤ 3	≤ 5	≤ 7	≤ 5	≤ 10
Stability (%)	≤ 12	≤ 40	≤ 12	$-$	≤ 10	$-$

Table. 2   Tentative specification for quality of recycled aggregate concrete

Kind of recycled aggregate concrete	Applications of recycled concrete	Used recycled coarse aggregate	Used fine aggregate
I	Reinforced concrete Plain concrete, etc.	1st class of recycled coarse aggregate	Normal aggregate
II	Plain concrete, etc.	2nd class of recycled coarse aggregate	Normal aggregate or 1st class of recycled fine aggregate
III	Subslab concrete	3rd class of recycled coarse aggregate	2nd class of recycled fine aggregate

3.3.2   Recycled aggregate

The reuse of demolished concrete as recycled aggregate is considered to be efficient. Tentative specifications for quality of both recycled aggregate and concrete made of recycled aggregate have been established by Japanese Ministry of Construction [4] (Tables 1 and 2).   The quality of recycled aggregate does not depend on the original concrete but is inferior to the normal aggregate [5].   As compared with normal aggregate, the absorption of recycled aggregate is generally greater than the normal aggregate because of the bonded mortar and cement paste.

If the recycled aggregate has low absorption after high-quality treatment, the strength and durability of the recycled aggregate concrete are of the same level as the normal concrete [6].   Obtaining high quality recycled aggregate is generally not easy due to the high cost and the problems in the reuse of the produced fine dust.   The quality of concrete made of low quality recycled aggregate is low.   In order to widen the use of the low quality concrete, both grading the concrete and matching the concrete to adequate application are important.

3.3.3   Completely recyclable concrete

The concept of the completely recyclable concrete (CR concrete) has been proposed by Tomosawa [7].   The CR concrete consists of cement, limestone coarse and fine aggregate and others, which are reusable as materials for manufacturing cement after demolishing the concrete.   In the future, when the amount of the demolished CR concrete exceeds the demand for manufacturing cement, the rest can be reused as recycled aggregate.   The concrete with this recycled aggregate from CR concrete is also CR concrete.   Pavement concrete is considered to be one of the potential applications of CR concrete.   The completely recyclable concrete has the following advantages :

- The limestone resources for cement can be stocked in the form of concrete.
- The emission of carbon dioxide (CO_2) in the manufacturing of cement can be reduced, when CR concrete is used as a raw material for cement.
- Slag aggregate and coal ash can be taken into the closed recycle loop.
- The by-produced fine dust can be reused as a material for cement.
- The sludge from ready mixed concrete can be also used as a raw material for cement.

3.4   Reduction of plywood form
The decrease of tropical forests is one of problems for the global environment.   It is reported that the forest area of 17,000,000 ha decreases every year in the developing countries [8].   As shown in Fig. 3, the tropical timber is used for fuel, commerce and exportation (3.9%).   Japan is the largest importer.   In Japan, two third of the imported tropical timber is consumed as plywood.   One fourth of the plywood is used as concrete form.

The timber for plywood must be cheap and strong and have no gnarls, annual rings and melting matter harmful to hydration of cement.   Such big trees widely range over the Philippines, Malay Peninsula and the Bornean islands.   In Japan, the timber is now imported mainly from Malaysia.   The export is prohibited in the Philippines.   The export tends to be prohibited in other countries.

The alternatives for the tropical timber plywood form are listed in Fig. 4.   The tropical timber plywood form would be replaced with the composite plywood form for the present and then replaced with non-timber forms in the future.   The composite

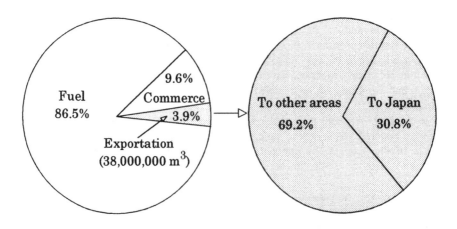

Fig. 3   Consumption of tropical timber

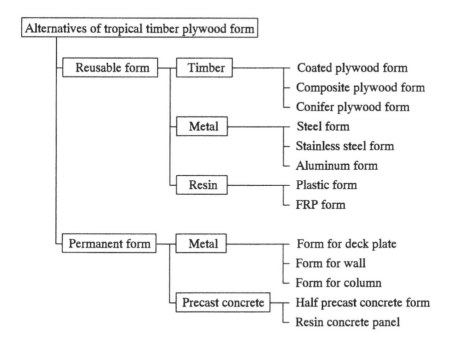

Fig. 4   Alternatives of plywood form made of tropical timber

plywood form consists of the tropical timber in both surfaces and the conifer in the center.   The consumption of tropical timber can be reduced by 70% in the composite form.   However, it should be noticed that the conifer grows slower than the tropical timber.

Permanent forms such as precast concrete forms and metal permanent forms are expected to be effective substitutions of the tropical plywood form.   There are several types of precast concrete forms.   Thin precast concrete forms are made with high strength concrete or polymer concrete having high flexural strength.   They have high durability.   Precast concrete form containing a set of reinforcement cage is often used to reduce the arrangement work of reinforcing bars.

In order to stimulate the use of precast concrete forms, the bond between precast concrete and cast-in-place concrete must be improved and the relation between the cost and the performance must be reasonable.   Standardization of precast concrete forms can reduce the cost.

3.5   Prolongation of life span of concrete structures
The prolongation of the life span of concrete structures is quite efficient to reduce construction wastes and therefore to restrain the consumption of resources.   The durability design, the inspection, the repair and the strengthening are all effective for the prolongation of the structural life span.   These subjects will be discussed in other papers in this workshop.

Concrete structures are often demolished not due to deterioration but rather of insufficient performance.   Thus, the ease of reformability of concrete structures is also important.

4   Conclusion

The environmental performance and the quality assurance are the new keywords in the design, construction and maintenance of concrete structures.   Most of the concrete technology can be related to the efficient use of resources.   The concept of the completely recyclable concrete proposed by Tomosawa is one of promising directions. To prevent the decrease of tropical forest, the use of permanent forms such as precast concrete forms is recommended.   The prolongation of the life span of concrete structures is quite efficient to restrain the consumption of resources.   The durability design, the maintenance, the rehabilitation and life-extension technique are all important for the prolongation.

5   References

1.   Ooi, T. (1994) Construction Wastes (Construction By-products) and Recycle. *Energy·Resource*, Vol. 15, No. 1, pp. 64-70, (Japanese).
2.   Maekawa, H (1992) Resource Countermeasure of Construction Waste in Japanese Ministry of Construction. *Journal of Japan Society of Civil Engineers*, Vol. 77, pp. 42-45, (Japanese).
3.   Yoshikane, T (1995) Example of Construction Using Concrete Waste. *Course on Use of Construction Waste, Chubu Branch of Japan Society of Civil Engineers*, pp. 24-43, (Japanese).
4.   Kawano, H (1994) Encouraging Reuse of Demolished Concrete by MOC Tentative Specification. *Cement Concrete*, No. 572, pp. 52-55, (Japanese).
5.   Morita, S et al. (1995) Basic Study on Recycled Concrete. *Proc. of the 8th Annual Conference of ZENNAMA*, pp. 61-66, (Japanese).
6.   Nanba, A et al. (1995) An Experiment for Improvement of Qualities of Recycled Concrete. *Proc. of the Japan Concrete Institute*, Vol. 17, No. 2, pp. 65-70, (Japanese).
7.   Tomosawa, F (1995) Towards Completely Recyclable Concrete. *Cement Concrete*, No. 578, pp. 1-8, (Japanese).
8.   Sakai, K (1994) Reduction of Tropical Forest and Alternative of Form Materials. *Journal of Japan Society of Civil Engineers*, Vol. 79-5, pp. 94-95, (Japanese).

22 THE STATE OF RE-USE OF DEMOLISHED CONCRETE IN JAPAN

H. KAWANO
Public Works Research Institute, Ministry of Construction,
Tsukuba, Japan

Abstract
This paper describes the state of re-use of demolished concrete in Japan. Firstly, it outlines the situation of demolished concrete under the Japanese Recycling Law and the amount produced including the re-use ratio over the last few years. It then introduces the activities of the Ministry of Construction of Japan to encourage its re-use. Some examples of pilot projects in which demolished concrete was used in various ways are shown. Technical problems concerning the further re-use of demolished concrete are also indicated.
Keywords: construction wastes, demolished concrete, recycle, recycled aggregate, re-use,

1 Introduction

Japanese construction work has been very active since the 1960's, as shown by the cement production in Japan in Table 1. Enormous amounts of infrastructure have been built, and investment in construction is still active, so the number of demolished structures is increasing. Until recently almost all demolished concrete was thrown away but there is now a shortage of dumping area.

Japanese aggregate production in 1994 was 900 million tons (Fig. 1). Some decades ago rivers were the main source but today the ratio of crushed aggregate is increasing. Furthermore, environmental problems concerning quarries and transportation roads have become more severe. It is therefore highly desirable to re-use demolished concrete

Integrated Design and Environmental Issues in Concrete Technology. Edited by K. Sakai. Published in 1996 by E & FN Spon, 2–6 Boundary Row, London, SE1 8HN, UK. ISBN 0 419 22180 8.

Table 1. Cement Production in Japan (million tons)

1950	1960	1970	1980	1990
6	24	58	87	84

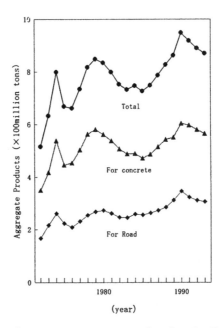

Fig. 1. Aggregate production in Japan

as an aggregate resource.

The word "recycle" is one of the most important keywords today. Recycling and saving resources will become more important in our daily lives as well as in many industrial activities in future. Because concrete is such an essential, mass-produced material like steel and soil in the construction industry, much effort has been made to recycle and conserve resources.

Technical developments on saving resources in concrete works started earlier than in other industries, such as the following:
- Utilization of by-products from other industries such as fly-ash, GGBFS, silica fume, gypsum
- Energy-saving in cement manufacturing, especially in the burning process
- Use of chemical admixtures

But the recycling of concrete, which means re-using concrete materials repeatedly, has not been achieved. It has been shown that if the re-use of demolished concrete is not increased, the Japanese construction industry will come to a deadlock in the near future.

Regardless of this fact, it is our obligation to re-use and conserve the resources of concrete materials in order to sustain the limited resources.

2 Laws on recycling

The Recycling Law, which encourages the recycling of almost every kind of material, has been in force since October 25, 1991 in Japan. The seven ministries of the Ministry of Construction, Ministry of Finance, Ministry of Transport, Ministry of Agriculture, Forestry and Fisheries, Ministry of International Trade and Industry, Ministry of Health and Welfare and the Environment Agency, drafted the law. Each ministry must designate specified industrial areas, by-products and waste materials, and show detailed methods of recycling these materials. The Ministry of Construction (MOC) designated the construction industry and specified excess construction soil, demolished asphalt concrete and demolished cement concrete as materials to be recycled.

The law requires each ministry to make a detailed plan. In compliance with the law the MOC is proceeding with the use of demolished concrete as a subbase material, foundation material and backfilling material.

The MOC and relevant organizations have drafted a plan (not a law) called "Recycle 21" which aims to recycle 100% of by-products. The plan anticipates that more than 90% of demolished concrete will be recycled by the year 2000.

3 The state of production and re-use of demolished concrete

Studies were conducted on the production of waste from construction works in 1990 and 1993. Figure 2 shows the production of such materials except soil material. The total amount and the ratio of each material were almost the same in 1990 and 1993. The total amount did not increase due to the economic recession, but the production of ready-mixed concrete in the last decade reached almost 400 million tons per year. The amount of demolished concrete accounts for only 6% of the annual concrete production, a figure which is expected to increase in the future.

Figure 3 shows the state of re-use of construction by-products. The ratio of re-use of demolished concrete increased from 48% in 1990 to 67% in 1993, and for asphalt concrete 50% to 78%. This indicates that the recycling law has been effective so far, but the re-use of demolished concrete is still limited to the subbase of road pavement.

The ratio of re-used wastes from housing works decreased during this period, probably because the housing industry suffered during the recession more severely than public works.

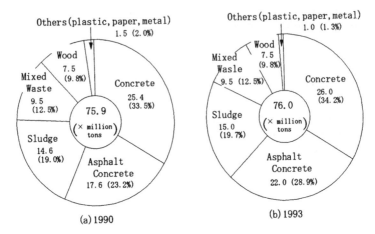

(a) 1990

(b) 1993

Fig. 2. Wastes from construction works

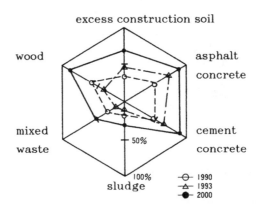

Fig. 3. Ratio of recycling

4 Method of encouraging the re-use of demolished concrete

Although research and development on the re-use of demolished concrete started around 1970 and the Ministry of Construction also conducted large-scale research projects on the re-use of waste materials from 1981 to 1985, in practice re-use did not increase significantly.

In 1991 a questionnaire was distributed asking why the ratio of re-use of demolished concrete had not risen. The main answers are summarized as follows:

· Lack of specifications or guidelines
· Concern about quality of recycled materials
· Incomplete supply system
· High cost for its quality

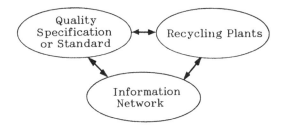

Fig. 4. Three requirements facilitating re-use

In order to solve these problems, the MOC will establish the three measures shown in Figure 4.

Firstly, the draft of tentative quality specifications for recycled concrete was issued and notified in April 1994 by the MOC. This draft is based on the research results of two large-scale research projects on the re-use of construction wastes, one from 1981 to 1985 and the other now being carried out since 1992. The specifications show three methods of re-use, for concrete aggregate, subbase materials of pavement and backfilling materials.

The MOC is considering establishing a system in which good recycling (crushing) plants register and are given priority to supply recycled materials to public works. This system is expected to foster better plants and to create a steady supply system.

Concerning the information network, we have developed a model on soil waste, which is currently working to connect the place of production and the place of re-use to facilitate re-use. For demolished concrete we must connect the place of production, recycling and re-use. From April 1995 we have been testing this network for demolished concrete in a limited area.

5 Examples of re-use of demolished concrete

It is not good to limit the scope of re-use of demolished concrete by regulations. Table 2 shows an example of re-usable structures. The upper part of the table shows the ways in which demolished concrete is used in the shape close to the original one, so little energy is needed to treat the demolished concrete but the scope of re-use is very limited. In the lower part of the table demolished concrete is crushed into small particles. In this way recycled materials can be used in many ways but higher energy is required. Therefore we have to choose the method of re-use considering the total conditions of cost, supply method, timing and so on.

Table 2. Example of Re-use

Demolished Member	Man-made Reef, Paving Stone
Broken into 20-40cm	Protection of Levee
Crushed (-50mm)	Subbase, Backfilling, Foundation Materials
Crushed and Worn (-40mm)	Concrete and Asphalt Concrete Aggregate Subbase Material, Backfilling Material
Powder (by-product through crushing)	Filler for Asphalt Concrete, Soil Stabilization Materials

The MOC is conducting many pilot projects on the re-use of demolished concrete.

6 Technical problems for further re-use of demolished concrete

Several technical problems remain to be solved for the further re-use of demolished concrete.

The MOC draft for the quality specifications of recycled materials are very much on the safe side, because there is not enough practical in-situ data on the long term characteristics such as durability. In order to encourage the re-use of recycled materials, the draft must be changed to a more rational one, balancing the required characteristics of structures using recycled materials.

The quality of recycled materials tends to vary more than normal materials, because usually the amount of demolished concrete from one site is small and the quality varies from site to site. The quality control methods for recycled materials must therefore be improved. For example, the method of testing the quality should be easier and quicker to enable more frequent testing.

Good quality recycled aggregate for durable concrete is more expensive than normal aggregate at present. The higher cost comes from its treatment procedure. When a conventional crusher is used, it takes a long time and much energy to get good quality aggregate. Therefore, low-cost, high-quality treatment machines need to be developed.

It has been shown that almost all recycled aggregate will be consumed as subbase material for road pavement in the next few years, yet we must enlarge the scope of re-use according to the increasing amount of demolished concrete, such as aggregate for asphalt concrete pavement, aggregate for filling concrete and so on.

It is almost impossible to obtain recycled aggregate free from impurities such as bricks, metals and plastics. Therefore, the effect of these materials on the structures using recycled materials must be assessed, and the contents of some of these materials may need to be limited. It is also necessary to develop a treatment procedure to eliminate impurities easily, including improving the machinery.

7 Conclusion

Besides the technical problems mentioned above, there are economic problems to be solved when recycling demolished concrete. In the early stage of the recycling, some administrative assistance is indispensable to accumulate technical data on many fields of re-use. Such research and development work on recycling and concrete technology should contribute to conservation of the global environment.

23 MECHANICAL PROPERTIES OF CONCRETES WITH NORTH CAROLINA RECYCLED AGGREGATES

S. H. AHMAD, D. FISHER and K. SACKETT
North Carolina State University, Raleigh, North Carolina, USA

Abstract
A study is underway at North Carolina State University (NCSU) to determine the effects of recycled fine aggregates on the properties of concrete. The objectives of the study are: (1) to determine the properties of the recycled aggregates typically used in construction of North Carolina pavements, (2) to determine the effects of different percent volume of recycled fine aggregate on the proportioning of the concrete mix and to develop an appropriate mix design utilizing recycled aggregates, and (3) to determine the effects of amount of recycled fines on the properties such as strength, stiffness and flexural strength of concretes made from these recycled aggregates, and to compare these to control concretes with natural aggregates.

This paper summarizes the results of the laboratory investigation of concretes made with use of recycled coarse and recycled fine aggregates. Concretes with various percentages of recycled fines and with 100% recycled coarse aggregates (RCA) were tested for strength, stiffness (modulus of elasticity), flexural strength, and split cylinder tensile strength. The results indicate that replacement of natural coarse aggregates with 100% recycled coarse aggregates reduced the compressive strengths in the order of 38% at the test age of 14 days. For concretes with 100% recycled coarse aggregates (RCA), increasing the percentages of recycled fine aggregates (RFA) decreased the compressive strength. The modulus of elasticity does not seem to show any definite trend. The flexural strength showed an initial decrease and then an increase as the percentage of RFA is increased beyond 25%. The split tensile strength seems to increase very slightly with the increase in the percentage of recycled fine aggregates.
Key Words: Flexural strength, modulus of elasticity, recycled aggregates strength, split tensile strength.

Integrated Design and Environmental Issues in Concrete Technology. Edited by K. Sakai. Published in 1996 by E & FN Spon, 2–6 Boundary Row, London, SE1 8HN, UK. ISBN 0 419 22180 8.

1 Introduction

With greater volumes and heavier vehicle weights on the highways, pavement distress is accelerating. There is a tremendous challenge to preserve the highway network by rehabilitating these roads to restore serviceability. There are a number of rehabilitation strategies that can be selected based on the condition of the existing pavement. In case of Portland Cement Concrete (PCC) pavements, rehabilitation may involve: (i) concrete pavement restoration, which would involve full depth and partial depth patching, slab stabilization, diamond grinding, joint and crack sealing, and drainage and shoulder repair, (ii) resurfacing, and (iii) reconstruction and/or recycling.

Several developments in the past ten years have made recycling more economical for all types of PCC pavements. These include development of improved equipment for breaking all types of PCC pavements (from plain to continuously reinforced), improvements in preparation for crushing operations, development of methods of steel removal to minimize hand labor, modification of crushing equipment to handle steel reinforcement.

There is a need for high quality construction materials, such as aggregate, to provide for the structures and transportation facilities for the future. However, existing supplies of natural aggregates are being depleted. There is a need to develop replacement sources of aggregates in areas where they are in short supply. Existing PCC pavements represent a readily available supply of aggregates that can be recycled for various uses.

A number of laboratory studies have been conducted [1-6] to determine the suitability of recycled aggregates for use in building new PCC pavements. Research [2,3,7] has shown an increase in freeze-thaw resistance of concrete made from recycled aggregates as compared to concrete using natural aggregates. Field studies have been conducted in several states [8-14]. Recycling projects in seven states (Michigan, Minnesota, Iowa, North Dakota, Wisconsin, Wyoming, and Oklahoma) were reviewed in an NCHRP synthesis [15].

A study is underway at North Carolina State University (NCSU) to determine the effects of recycled fine aggregates on the properties of concrete. The objectives of the study are: (1) to determine the properties of the recycled aggregates typically used in construction of North Carolina pavements, (2) to determine the effects of different percent volume of recycled fine aggregates on the proportioning of the concrete mix and to develop an appropriate mix design utilizing recycled aggregates, and (3) to determine the effects of amount of recycled fines on the properties such as strength, stiffness, and flexural strength of concretes made from these recycled aggregates, and to compare these to control concretes with natural aggregates.

2 Sources of recycled aggregates

The concrete slab for the crushing and obtaining the recycled aggregates was identified by the research team in collaboration with the pavement management unit of the North Carolina Department of Transportation (NCDOT). The slab was obtained from Davie County, and the material was removed from Interstate 40, Milepost 165, Station 675+00, in the east bound lane. NCDOT information states that Smithgrove was the source of the course aggregate and Yadkin River Natural Sand along with North Quarry Manufactured Sand provided the fine aggregates used in the original concrete slab mix.

A small section of the slab, 365x 244 cm in size, was transported to the NCDOT Materials and Tests Unit for evaluation. The slab was cored, and tests were conducted to determine the chloride content at various depths of the slab. The results of these tests are shown in Table 1. The maximum allowable chloride content is 1.2 kg/m^3. In addition, compression and split cylinder tests were conducted on cores from the slab. The cores for these tests were about 150 mm in length with a diameter of about 80 mm. Results of the tests on the cores are shown in Table 2.

Table 1. Chloride content test results on concrete slab

Sample No.	Depth (mm)	Cl %	kg/m^3
1A	25	0.027	0.63
1B	50	0.018	0.41
1C	75	Trace	< 0.36
1D	100	Trace	< 0.36
1E	126	Trace	< 0.36
2A	25	0.020	0.46
2B	50	Trace	< 0.36
2C	75	0.046	1.12
2D	100	0.025	0.58
2E	126	0.025	0.58
2F	152	0.025	0.58

Table 2. Material properties of concrete slab

Sample No.	Test Type	Strength (MPa)
1	Split Tensile	4.00
2	Compression	5.01
3	Compression	44.70
4	Compression	49.30

The concrete slab was crushed in the Phoenix Recycler located in the Havelock region to determine the percentage of fines generated for the range of cement contents normally utilized in North Carolina. It was intended that this would approximate a field crushing operation as closely as possible. The crusher products were sieved to obtain coarse and fine fraction grading test results. The maximum size of the crushed aggregate was 38 mm. The coarse and fine fraction of the recycled PCC aggregate were stockpiled separately to avoid segregation problems.

3 Laboratory tests for physical properties

Following the sieving operation, a series of tests were performed on the recycled aggregates. For the coarse aggregate (material retained on the 4.75 mm sieve), testing was done for bulk specific gravity, percent absorption, and unit weight. Additionally, the abrasion resistance tests were conducted. Percent absorption, specific gravity, and fineness modulus tests were performed on the fine aggregate. All tests were performed according to ASTM specifications. The results of these tests are shown in Table 3.

Table 3. Physical properties of recycled aggregates

Recycled fine aggregate (RFA)

Bulk specific gravity	2.39
Bulk specific gravity (SSD)	2.67
% Absorption	10.5%
Fineness modulus	2.58

Recycled coarse aggregates (RCA)

Dry rodded unit weight	1477 kg/m^3
Bulk specific gravity	2.46
Bulk specific gravity (SSD)	2.61
Apparent specific gravity	2.89
% Absorption	6.1%
LA Abrasion (Grade-A)	30.1%

4 Laboratory tests for mechanical properties

The variables of the test program were the percentage of recycled fines, the type of testing parameter, and the age at testing. The percentage of the recycled fine aggregate (RFA) was varied between 0% to 100% for concretes with 100% recycled coarse aggregates (RCA). Tests were conducted for compressive strength, elastic modulus, flexural strength, and split tensile strength. The tests at the ages of 14 days and 28 days have been conducted, and tests at the age of 9 months will be conducted in the near future.

4.1 Mixture proportions

The preliminary mix proportions used were similar to that used by the NCDOT with a cement factor of 312 kg/m^3. This mix design produces a concrete that meets the target flexural strength of 3.8 MPa at the age of 14 days, which is the minimum required flexural strength for pavements. The mix design was adjusted slightly for producing a concrete with slightly higher flexural strength at 14 days, since a drop in strength was anticipated upon substituting the recycled aggregates. The adjusted mix design has a cement content of 367 kg/m^3, while the water/cement ratio was kept constant. This mix design is termed as "control" mix and designated as NCSU Mix 1. Two batches were made, the first using all natural aggregates, the second replacing 100% of the natural coarse aggregate with recycled coarse aggregate (RCA). The recycled aggregates were not washed prior to use. For all batches, the target slump was 5 cm ± 2.5 cm, and the target air content was 5% ± 1.5%. The proportions of the constituent materials used in these batches are shown in Table 4. The air entraining agent used was neutral vinsol resin with 17% solids, and a napthalene based high range water reducer (HRWR) was used.

Table 4. Dry weights for trial batches

Batch identification	Material	Material amounts (kg/m^3)
NCDOT Mix Cement - 312 kg/m^3 0% RCA, 0% RFA	Cement FA CA Water Air entraining agent (AEA) High range water reducer (HRWR)	312 722 1112 148 173 mL 1194 mL
NCSU Mix 1 Cement - 367 kg/m^3 0% RCA, 0% RFA	Cement FA CA Water Air entraining agent (AEA) High range water reducer (HRWR)	367 705 1086 173 173 mL 3475 mL
NCSU Mix 2 Cement - 367 kg/m^3 100% RCA, 0% RFA	Cement FA RCA Water Air entraining agent (AEA) High range water reducer (HRWR)	367 646 1005 173 173 mL 1307 mL
NCSU Mix 3 Cement - 367 kg/m^3 60% RCA, 0% RFA	Cement FA RCA CA Water Air entraining agent (AEA) High range water reducer (HRWR)	367 646 603 432 173 173 mL 1307 mL

4.2 Mechanical properties

Following these preliminary batches, the substitution of the natural fines with recycled fines was initiated. A percentage of the natural fines was replaced volumetrically with recycled fine material. Batches were made by replacing 25, 50, and 75 percent of the natural fine aggregate with recycled fine aggregate. A batch using 100 percent recycled fines was also attempted, but it failed to attain the required air content.

Tests for compression, stiffness (modulus of elasticity), and split tensile strength were performed on 100 mm x 200 mm cylinders. Flexural strength tests were conducted on 100 mm x 100 mm x 356 mm beam specimens. For each test variable, three replicate specimens were tested. Table 5a shows the test results for strength and modulus of elasticity at different test ages.

Table 5a. Summary of laboratory test results for strength and modulus of elasticity

Cement factor	Test type	% RCA	% RFA	Test age: 14 days			Test age: 28 days			Test age: 9 months		
NCSU Mix 1	Comp.	0%	0%	50.1	52.2	50.7	49.6	56.5	53.5	C*	C	C
NCSU Mix 2	Str.	100%	0%	32.3	30.8	31.2	33.1	37.6	35.9	C	C	C
367 kg/m^3	f_c	100%	25%	27.8	30.1	29.9	X	X	X	* specimen		
	(100x200mm)	100%	50%	27.6	29.4	29.4	X	X	X	casted for		
		100%	75%	28.5	27.6	28.3	X	X	X	testing		
		100%	100%									
NCSU Mix 3	Comp.	60%	0%							C	C	C
367 kg/m^3	Str.	60%	25%							C	C	C
	f_c	60%	50%							C	C	C
	(100x200mm)	60%	75%							C	C	C
		60%	100%									
NCSU Mix 1	El. Mod.	0%	0%	28.7	29.8	31.3	34.4	40.3	34.1	C	C	C
NCSU Mix 2	E (x10^3 MPa)	100%	0%	28.1	25.2	26.8	---	23.2	21.2	C	C	C
367 kg/m^3	(100x200mm)	100%	25%	23.0	23.0	24.3	X	X	X	X	X	X
		100%	50%	33.1	---	26.7	X	X	X	X	X	X
		100%	75%	23.8	24.3	23.9	X	X	X	X	X	X
		100%	100%									
NCSU Mix 3	El. Mod.	60%	0%							C	C	C
367 kg/m^3	E (x10^3 MPa)	60%	25%							C	C	C
	(100x200mm)	60%	50%							C	C	C
		60%	75%							C	C	C
		60%	100%									
NCDOT	Comp. Str.	0%	0%	30.3	29.2		38.4	29.7	34.4	C	C	C
Mix	Modulus	0%	0%	33.9			27.6	21.8	25.6	C	C	C
312 Kg/m3	(x103^3)	0%	0%	-----	26.8		4.00	3.96	4.14	C	C	C
	Flex. Strength	0%	0%	24.1			3.59	3.00	2.93	C	C	C
	Split Cylinder			3.55	4.07							
				4.03								
				3.41	2.01							
				3.07								

The replacement of the natural coarse aggregate with 100% RCA reduces the average compressive strength of concrete at 28 days from 53.2 MPa to 35.5 MPa, which is a reduction of 33%. For the test age of 14 days, the strength reduces from 51.0 MPa to 31.4 MPa, which is a reduction of 38%.

The effect of increasing the percentage of recycled fine aggregate (RFA) on the strength and stiffness of concrete with 100 recycled coarse aggregate (RCA) is shown in Figures 1-2. Increasing the percentage of recycled fine aggregate (RFA) in concretes with 100% recycled coarse aggregate (RCA) decreases the compressive strength of concrete at the test age of 14 days. The modulus of elasticity does not seem to show any definite trend with increasing the percentage of RCA. First it decreases when the RFA is 25%, but then it shows an increase when the percentage of RFA is 50%.

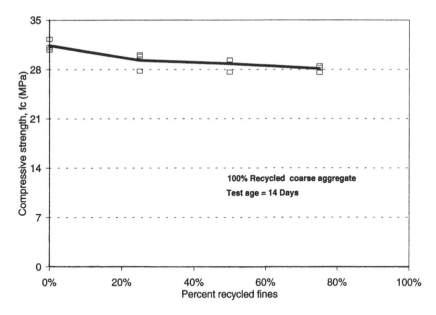

Fig. 1 Effect of recycled fines on compressive strength of concrete

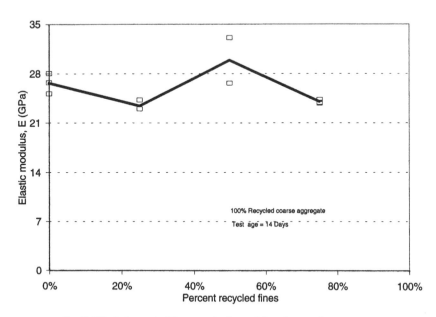

Fig. 2 Effect of recycled fines on elastic modulus of concrete

Table 5b shows the test results for flexural strength and split cylinder strength for concretes with recycled aggregates. These results are for different tests ages.

Table 5b. Summary of laboratory test results for flexural and split cylinder strength

Cement factor	Test type	% RCA	% RFA	Test age: 14 days			Test age: 28 days			Test age: 9 months		
NCSU Mix 1	Flexural	0%	0%	5.55	5.17	---	5.90	5.48	4.62	C	C	C
	strength	100%	0%	3.93	3.96	3.93	4.14	4.17	4.14	C	C	C
NCSU Mix 2	f_r	100%	25%	2.93	3.55	3.17	X	X	X	X	X	X
367 kg/m^3	(100x100x	100%	50%				X	X	X	X	X	X
	356mm)	100%	75%	3.52	3.59	3.28	X	X	X	X	X	X
		100%	100%	3.34	3.52	3.10						
NCSU Mix 3	Flexural	60%	0%									
367 kg/m^3	strength	60%	25%							C	C	C
	f_r	60%	50%							C	C	C
	(100x100x	80%	75%							C	C	C
	356mm)	60%	75%	3.69	3.96	4.00				X	X	X
		60%	100%	4.03	4.14	4.10	X	X	X	C	C	C
NCSU Mix 1	Split cylinder	0%	0%	3.14	2.90	3.79	4.07	4.65	4.10	C	C	C
NCSU Mix 2	f_{ct}	100%	0%	2.79	2.48	2.86	3.10	2.86	3.34	C	C	C
367 kg/m^3	(100x200mm)	100%	25%	3.10	2.65	3.03	X	X	X	X	X	X
		100%	50%	3.10	2.79	2.86	X	X	X	X	X	X
		100%	75%	3.17	2.86	2.93	X	X	X	X	X	X
		100%	100%									
NCSU Mix 3	El. mod.	60%	0%							C	C	C
367 kg/m^3	f_{ct}	60%	25%							C	C	C
	(100x200mm)	60%	50%							C	C	C
		60%	75%							C	C	C
		60%	100%									
NCDOT	Comp. str.	0%	0%	30.3	29.2		38.4	29.7	34.4	C	C	C
Mix	Mod. (x10^{33})	0%	0%	33.9			27.6	21.8	25.6	C	C	C
312 Kg/m3	Flex. strength	0%	0%	-----	26.8		4.00	3.96	4.14	C	C	C
	Split cylinder	0%	0%	24.1			3.59	3.00	2.93	C	C	C
				3.55	4.07							
				4.03								
				3.41	2.01							
				3.07								

The effect of increasing the percentage of recycled fine aggregate (RFA) on the flexural strength and split cylinder strength of concrete with 100% recycled coarse aggregate (RCA) is shown in Figures 3-4. Overall, the 14 day flexural strength decreases from 3.94 MPa to 3.32 MPa, a decrease of 16%. The split tensile strength seems to increase very slightly with the increase in percentage of recycled fines. From Figure 3, it can be seen that flexural strengths at test age of 14 days for concretes with 100% RCA and various percentages of RFA are lower than the required minimum strength of 3.8 MPa. The most significant reduction in strength appears to be from the substitution of RCA for natural coarse aggregate, while increasing percentages of RFA appeared to have very little detrimental effect on strength (Table 5a).

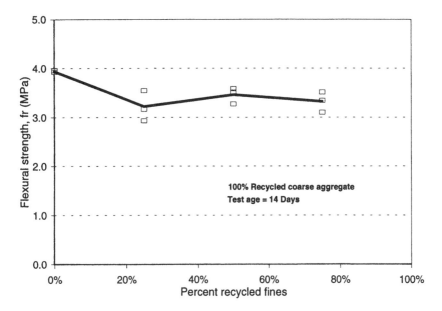

Fig. 3. Effect of recycled fines on flexural strength of concrete

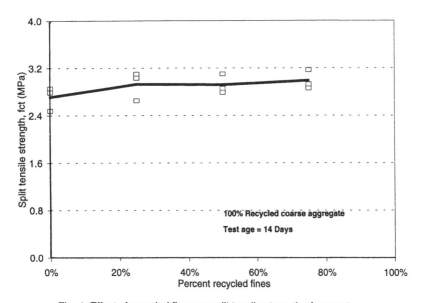

Fig. 4 Effect of recycled fines on split tensile strength of concrete

Preliminary testing has been completed using 75% recycled fine aggregates with 60% and 80% recycled coarse aggregates. These results are included in Table 5a and 5b. It appears that mixes with 60% replacement of recycled coarse aggregates for natural coarse aggregates are expected to meet the minimum target flexural strength of 3.8 MPa at the age of 14 days. Tests are in progress on NCSU Mix 3 using 60% recycled coarse aggregate and varying percentages of recycled fine aggregates.

5 Summary

In this study, experimental investigation was conducted to study the effect of recycled fines on selected mechanical properties of concrete. The test variables included the type of tests, percentages of recycled fine aggregates and the age at testing.

The results indicate that replacement of natural coarse aggregates with 100% recycled coarse aggregate reduced the compressive strengths in the order of 38% at the test age of 14 days.

For concretes with 100% recycled coarse aggregates, increasing the percentages of recycled fine aggregates decreases the compressive strength. The modulus of elasticity does not seem to show any definite trend. The flexural strength has an overall decrease of 16%. The split tensile strength seems to increase very slightly with the increase in the percentage of recycled fine aggregates.

6   Acknowledgments

The support of the Materials and Tests Unit of North Carolina Department of Transportation is gratefully acknowledged.

7   References

1.   Epps, J.A. and O'Neal, R. J. (1975) Engineering, Economy and Energy Considerations Recycling Pavement Materials, Cooperative Research Project 214-12, Texas Transportation Institute and Texas State Department of Highways.
2.   Buck, A.D. Recycled Concrete, Miscellaneous Paper C72-14, U.S. Army Engineer Waterways Experiment Station, C.E., Vicksburg, Mississippi.
3.   Buck, A.D., Recycled Concrete, Miscellaneous Paper C72-14, Report 2, additional investigation, U.S. Army Engineer Waterways Experiment Station, C.E., Vicksburg, Mississippi.
4.   U.S. Army Engineer Waterways Experiment Station, C.E., (1949) *Handbook of Concrete and Cement,* Vicksburg, Mississippi.
5.   Bergren, J.V. and Britson, R.A. (1977) Portland Cement Concrete Utilizing Recycled Pavement, Iowa Department of Transportation, Division of Highways.
6.   Frondistou-Yannas, S. and Itoh, T. (1977) Economic Feasibility of Concrete Recycling, *Proceedings*, ASCE, Vol. 103, No. St4, pp. 885-899.
7.   Fergus, J.S.   Laboratory Investigation and Mix Proportions for Utilizing Recycled Portland Cement Concrete as Aggregate, paper on research performed in partial fulfillment of the requirement for a doctoral degree at Michigan State University.
8.   Halverson, A.D., (1985) Recycling Portland Cement Concrete Pavement, Minnesota Department of Transportation, Office of Research and Development.
9.   McCarthy, G.J.   (1985) Recycling of Concrete Freeways by Michigan Department of Transportation, in *Transportation Research Record 1040: Concrete Pavement Construction,* Transportation Research Board, National Research Council, Washington, D.C. pp. 21-24.

10. McCarthy, G.J. and McCreery, W.J. (1985) Michigan Department of Transportation Recycles Concrete Freeways, *Proceedings of the 3rd International Conference on Concrete Pavement Design and Rehabilitation,* Purdue University pp. 643-647.
11. Bancroft, K.S., Interoffice letter to Fred Copple, Summary of 1985 Concrete Pavement Recycling Projects near Albion and Kalamazoo, Michigan, Research Project 78B-99, Michigan Department of Transportation.
12. Gendreau, C.J. Recycling Concrete Pavement in North Dakota, Internal report, North Dakota State Highway Department.
13. Swedeen, K.J. (December 1985) Recycled Portland Cement Concrete Pavement in Wyoming, presented at the 1985 Asphalt/Concrete Congress in Casper, Wyoming.
14. Hankins, R.B. and Borg, T.M. (1986) Recycling PCC Roadways in Oklahoma, in *Transportation Research Record 986: Construction: Quality Control and Specifications,* Transportation Research Board, National Research Council, Washington, DC, pp. 1-4.
15. Recycling of Portland Cement Concrete Pavements, (1989) NCHRP Synthesis of Highway Practice 154, Transportation Research Board, National Research Council, Washington, DC.

24 TOWARDS COMPLETELY RECYCLABLE CONCRETE

F. TOMOSAWA and T. NOGUCHI
University of Tokyo, Tokyo, Japan

Abstract

The problems caused by concrete are waste disposal, limestone exhaustion, exhaustion of aggregate resources, and environmental pollution and nature destruction due to waste concrete disposal and aggregate excavation as well as carbon dioxide emission from cement production, which require decisive solutions from the standpoint of the global environment. The best way to solve these problems is to make the total amount of concrete reusable over and over again as a structural material like steel and aluminum – to make it completely recyclable. This paper describes the current problems regarding concrete and how they can be resolved by making concrete completely recyclable. Experiments to prove complete recyclability of concrete and simulation to forecast the future situations of concrete production were conducted.

Keywords: recycle, waste concrete, ecological concrete, cement production, limestone resource, industrial byproduct, CO_2 emission, global environment

1 Foreword

Concrete is an excellent material considered to be forever useful for structures and infrastructures. A glaring defect of concrete, however, is that it is very difficult to recycle. The problem of disposing of concrete mass as industrial waste that results from the stock of enormous amounts of concrete structures at present and to be produced in future is beginning to be identified and addressed as an acute problem, as well as the problem of expiring limestone resources.

Integrated Design and Environmental Issues in Concrete Technology. Edited by K. Sakai. Published in 1996 by E & FN Spon, 2–6 Boundary Row, London, SE1 8HN, UK. ISBN 0 419 22180 8.

Uses for waste concrete mass are being sought for by various institutions, but currently the possibilities center on road bed materials and recycled aggregate. These are far from sufficient, leaving numerous problems unsolved. The problems caused by concrete being difficult to recycle are not limited to waste disposal, limestone exhaustion, and exhaustion of aggregate resources. Environmental pollution and nature destruction due to waste concrete disposal and aggregate excavation as well as carbon dioxide emission from cement production are posing pressing problems that require decisive solutions from the standpoint of the global environment.

The best way to solve these problems is to make the total amount of concrete reusable over and over again as a structural material like steel and aluminum – to make it completely recyclable. This has been deemed impossible, but can be made possible with a very simple but radical change of conception[1].

This paper describes the current problems regarding concrete and how they can be resolved by making concrete completely recyclable.

2 Current problems regarding recycling of concrete

Concrete is a material difficult to recycle, and once disposed of, currently most of it has to be dumped. This chapter discusses the problems that may result when the current situation is left untouched.

2.1 Waste disposal problem

The annual concrete waste (understood quantitatively) in Japan is currently 25 million t, of which 10 million t are reported to be used for road beds and the rest dumped[2]. However, the estimated concrete input to the construction sector suggests a much larger amount of waste concrete. The concrete waste from removed buildings alone, for instance, is estimated to be 71 million t and 110 million t in 1995 and 2021, respectively, on an input basis from the statistics of removed buildings and original units of concrete used as classified by type of structure[2]. This poses the question of whether waste concrete is, even at present, appropriately disposed of. The increasing waste will inevitably lead to a serious shortage of dumping sites and environmental pollution.

A number of problems were pointed out regarding seismic measures after the Hanshin Awaji Earthquake of January 17 this year, including the generation of an enormous amount of debris and its disposal. Structures should, from the outset, be constructed so as not to be damaged so severely as to require disposal. If concrete was a completely recyclable material, this problem would not have been this serious.

2.2 Problem of recycling

Recycling of waste concrete is being investigated in various institutions. The largest part (approx. 10 million t) of recycled waste concrete is for road beds, and the applicability of waste concrete for road beds is nearly established. The demand for crushed stone for road beds is approximately 300 million t, which is sufficient.

Regarding recycled concrete aggregates, the Ministry of Construction issued a quality standard for civil engineering structures in March last year, and plans to propose quality and usage standards for buildings in March this year. However, recycled aggregate produced by crushing waste concrete still has quality–related problems, and its improvement involves the problems of cost and material balance regarding the treatment of fines.

In other words, utilization of waste concrete as recycled aggregate by crushing it is being investigated, but sufficient milling is required to obtain recycled aggregate of good quality. This process, however, produces a large amount of dust, which is difficult to recycle or dispose of, and this hampers the production and utilization of recycled aggregate.

2.3 Problem of aggregate resources

The total annual demand for concrete aggregate in Japan is approximately 400 million t. Though the problem of aggregate resources has been posed and investigated since the 1950s, the solutions applied so far are not satisfactory. The maintenance of aggregate supply is threatened by the shortage of resources and environmental destruction induced by aggregate excavation.

2.4 Problem of limestone resources

The Japanese limestone reserve is estimated to be approximately 58 billion t, of which minable crude ore is approximately 38 billion t, and the proved minable ore is 9.8 billion t[3].

Currently 200 million t are mined every year, and 15% of it, or 30 million t, is used as concrete aggregate. Simple calculation indicates that the proved minable ore will be used up in 50 years.

2.5 Problem of CO_2 emission

Cement production using limestone as the main material releases approximately 0.5 t of carbon dioxide fixed by limestone per ton of cement. As long as limestone is used as the main material for cement, this original unit of emission cannot be reduced. The CO_2 emission from the cement industry ranks second (20%) among Japanese industries, accounting for 5.7% of the total CO_2 emission[4].

2.6 Utilization and recycling of industrial waste

Industrial byproducts, such as blast–furnace slag, fly ash, and products derived from these byproducts, are utilized as materials for cement and concrete. They seem to have been effectively utilized, but they are eventually disposed of when the concrete is disposed of, never to be reused.

As stated above, future recycling and waste disposal have not been sufficiently considered for concrete, when 200 million m^3 of it is annually used in Japan.

3 What is completely recyclable concrete?

The discussion of concrete recycling has centered on crushing waste concrete for use as aggregate, without ever considering the recycling of waste concrete back to a raw material state, as in the case of steel, aluminum, or paper. On the other hand, no one would think that steel and aluminum can be recycled simply by crushing or cutting. Why should recycling of concrete be discussed only in the realm of it remaining in the form of concrete as such? Can it not be returned to cement by calcining in a kiln? This shift of conception easily leads to the concept of completely recyclable concrete.

Concrete referred to as completely recyclable concrete here is defined as follows:

Completely recyclable concrete is a concrete or mortar, whose binders, additives, and aggregates are all made of cement or materials of cement and all of these materials can be used as cement materials or recycled aggregate after hardening. A concrete containing recycled aggregate from a completely recyclable concrete is also a completely recyclable concrete. In this way, such a concrete can be recycled endlessly.

The concept and its basic pattern are simply displayed in Fig. 1[5].

4 Experiment to prove complete recyclability

As a most basic example, a concrete containing ordinary portland cement, crushed limestone and crushed limestone dust is produced. This concrete is then crushed into a blended raw cement material with the required adjustment of the components. This material is calcined in an electric furnace and the resulting clinker is added with a suitable amount of gypsum, and milled to produce recycled cement. Concrete is produced again with this cement using limestone aggregates.

The materials used are as follows:

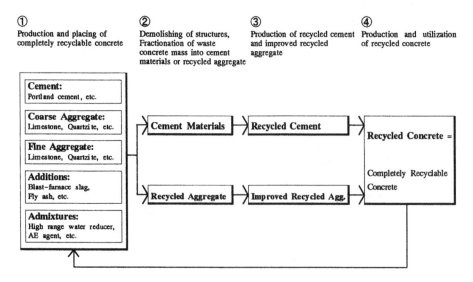

Fig. 1 The concept and basic pattern of completely recyclable concrete

- *Cement* : ordinary portland cement with a specific gravity of 3.16.
- *Coarse aggregate* : crushed limestone containing 54.2% CaO. The oven-dry specific gravity is 2.62.
- *Fine aggregate* : crushed limestone dust containing 54.2% CaO. The oven-dry specific gravity is 2.61.
- *Chemical admixtures* : an air-entraining and water-reducing agent and an air-entraining and high-range water-reducing agent.

The mix proportions of the original concrete are as given in Table 1. "Normal strength concrete" is concrete normally used for buildings, and "high strength concrete" is that used for such structures as high-rise reinforced concrete buildings.

The properties of the concrete produced are as given in Table 2. The chemical analysis values for the normal strength concrete are as given in Table 3. The ignition loss, 42.92%, includes free water, combined water, and carbon dioxide within limestone aggregate in the concrete. With this being excluded, calcium oxide is excessive and silicon dioxide, aluminum oxide, and ferric oxide are insufficient in the original concrete as a cement material. These components as well as sodium hydroxide, potassium hydroxide and sulfuric acid are therefore added to coarsely crushed original concrete. The mixture is then finely milled into a blended material for cement. This is then calcined in an electric furnace, added with gypsum, and ground using a test ball mill, to produce recycled cement. The chemical composition and physical test results are given in Tables 4 and 5, respectively. The strength of the recycled cement was the same or higher than cements on the market, and the setting time was also within the range of that of ordinary cement.

Table 1. Mix proportions of original concrete

Type	Grade	Slump	W/C	Air	Unit Weight (kg/m3)				Admixture
	(N/mm2)	(cm)	(%)	(%)	Water	Cement	Fine Agg.	Coarse Agg.	(ml/m3)
Normal	24	18	58	4	184	320	732	1048	805
High Strength	60	21	30	2	171	571	600	1057	4100

Table 2. Mechanical properties of original concrete

Type	Grade (N/mm2)	Compressive Strength at 28 Days (N/mm2)	Youngs Modulus at 28 Days (N/mm2)
Normal	24	31.6	39100
High Strength	60	67.6	48000

Table 3. Chemical analysis values for original normal strength concrete (JIS R 5202)

ig.loss	SiO2	Al2O3	Fe2O3	CaO	MgO	SO3	Na2O	K2O	TiO2	MnO	P2O3
42.92	4.61	0.78	0.48	49.83	0.76	0.23	0.07	0.08	0.05	0.03	0.06

99.88 in all

Table 4. Chemical analysis values for recycled cement (JIS R 5202)

ig.loss	SiO2	Al2O3	Fe2O3	CaO	MgO	SO3	Na2O	K2O	TiO2	MnO	P2O3
0.88	21.18	4.98	2.75	66.23	1.02	1.89	0.19	0.24	0.10	0.05	0.08

insol.=0.21, 99.81 in all

Table 5. Physical test results of recycled cement (JIS R 5201)

Specific Gravity	Specific Surface Area (cm2/g)	Remains on 90 μ sieve (%)	Setting Water (%)	Initial h–m	Final h–m	Sound– ness	Flow (mm)	Strength (N/mm2) Compression 3d	7d	28d	Flexure 3d	7d	28d
3.13	3340	1.8	26.6	2–00	2–50	good	233	15.5	25.3	44.1	3.67	5.51	7.24

Table 6. Mix proportions of recycled concrete

Type	Grade (N/mm2)	Slump (cm)	W/C (%)	Air (%)	Unit Weight (kg/m3) Water	Cement	Fine Agg.	Coarse Agg.	Admixture (ml/m3)
Normal	24	18	58	4	170	296	862	971	805
High Strength	60	21	30	2	171	571	600	1057	4100

Table 7. Mechanical properties of recycled concrete

Type	Grade (N/mm2)	Compressive Strength at 28 Days (N/mm2)	Youngs Modulus at 28 Days (N/mm2)
Normal	24	35.2	39000
High Strength	60	66.8	46500

Two types of concrete are mixed with mix proportions given in Table 6 using the recycled cement in combination with the same aggregate as used in the original concrete, to measure the compressive strength and Youngs modulus. The results are given in Table 7. The comparison of Table 7 with Table 2 reveal that the compressive strength and Youngs modulus of the recycled concretes are equivalent with those of the original concretes.

Consequently, it is confirmed that the entire amount of the original concrete can be completely recycled as a cement material.

5 Solving problems by using completely recyclable concrete

Completely recyclable concrete allows concrete, previously considered nonrecyclable, to be a completely recyclable material. This will provide fundamental solutions to the present and future problems regarding concrete as stated below.

5.1 Waste disposal problem

Use of completely recyclable concrete eliminates waste disposal when concrete structures and products are disposed of, as all of the waste concrete can be used as a cement material.

All sludge and returned concrete at ready–mixed concrete plants can also be used as cement materials.

5.2 Recycled aggregate

Part of completely recyclable concrete waste is utilized as recycled aggregate. This can be used in combination with aggregates used for initial completely recyclable concrete. Concrete containing this recycled aggregate will be another completely recyclable concrete, and the recycling continues endlessly as this is repeated.

Dust produced when producing recycled aggregate can be used as a cement material. Sufficient milling produces recycled aggregate of good quality.

5.3 Aggregate availability problem

By the use of completely recyclable concrete, environmental destruction due to scattered aggregate excavation on a small scale can be avoided, and crushed limestone, with which environmental protection is easier, will be used actively. It will also promote the use of various metallurgical aggregates and fly ash. Advanced treatment techniques will produce recycled aggregates of good quality, and the repeated use of recycled aggregate will alleviate the constraint on aggregate availability. Consequently, most of the problems related to aggregates, such as exhaustion of aggregate resources and environmental destruction due to aggregate excavation will be solved or slackened.

5.4 Limestone availability problem

Limestone resources are permanently stored and accumulated in the form of structures, eliminating the problem of resource consumption. This will resolve the future problem of limestone availability.

5.5 CO_2 emission problem

The use of already calcined clacareous materials as part of cement materials reduces the original unit of CO_2 emission at the time of cement production. The reduction in the cement calcination energy will lead to a reduction of CO_2 emission in terms of energy as well, which accounts for 42% of the total CO_2 emission, alleviating the impact of CO_2 emission by cement production on the environment.

This tendency toward reduction can be sharpened by allocating the hardened cement portions of completely recyclable concrete to the cement materials.

5.6 Utilization and recycling of industrial waste

Industrial byproducts, such as blast–furnace slag, fly ash, and materials produced from them, will be actively utilized as materials for cement and concrete. They can also be endlessly recycled by being used for completely recyclable concrete. Limestone used for the formation of blast–furnace slag can be recycled as well.

As stated above, completely recyclable concrete can be recycled again and again like steel and aluminum, and can have profound effects as a measure against problems concerning the global environment and resources. It will provide a fundamental solution to the future problem of concrete waste dumping; radically improve the present and future aggregate availability; allow permanent storage of limestone resources; inhibit CO_2 emission from cement production; and promote the utilization of various industrial byproducts.

6 Future prospects

The future situations of concrete production were simulated by assuming that ordinary concrete is replaced by completely recyclable concrete containing crushed limestone and limestone dust year after year and that the annual growth rate of production is 5%. The results are shown in Figs 2 to 5.

The following conditions were adopted:

1. The limestone amount required for producing 1 t of ordinary cement = 1.1 t
2. The recovery of demolished concrete = 95%
3. When completely recyclable concrete or "recycled" completely recyclable concrete is demolished, initially the total amount will be used as cement materials. After such materials have become oversupplied, they will be recycled as recycled aggregates.

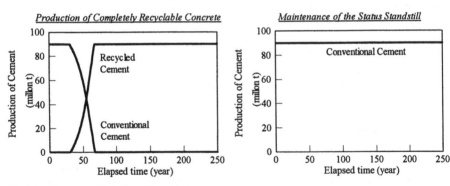

Fig. 2 Change of produced cement

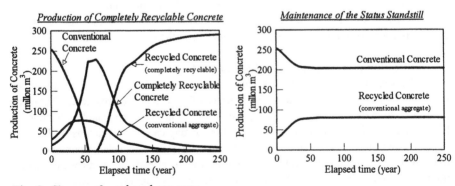

Fig. 3 Change of produced concrete

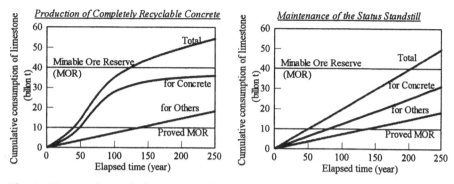

Fig. 4 Change of cumulative consumption of limestone

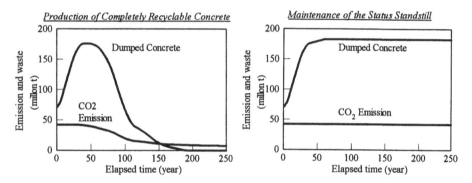

Fig. 5 Change of CO2 emission and dumped concrete

4. Structures will be evenly demolished between 31 and 60 years after construction.

5. In 25 years, 10%, 50%, and 40% of demolished conventional concrete will be used as silica materials for cement, road bed materials, and recycled aggregate, respectively.

6. Completely recyclable concrete (limestone aggregate type) will increase by 5% every year until the rest of the demand can be supplemented by recycled concrete.

7. Conventional concrete remaining after its production is discontinued, as well as its recycled concrete, will be used as road bed materials.

The simulation revealed the following:

1. Conventional concrete will cease to be produced in 50 years. After 70 years, concrete recycled from completely recyclable concrete will be produced.

2. The cement materials will shift from limestone to waste concrete in 30 to 70 years.

3. The limestone consumption will reach the proved minable ore reserve in 40 years plus even if no changes are made. If completely recyclable concrete is used, the proved minable ore reserve will be reached slightly earlier, requiring an increase

in the minable ore, but then limestone will be recycled.

4. Dumped concrete will reach its peak in 50 years, and then shift to a downward trend.

5. The CO_2 emission will gradually start to diminish in around 30 years.

7 Afterword

A few questions remain to be solved to implement the real use of completely recyclable concrete. However, the spread of this concrete will enable the entire amount of demolished concrete to be used as cement materials and recycled aggregate for completely recyclable concrete in the future. Cooperation among the parties concerned will be necessary to have this concrete widely accepted. It is also necessary to investigate measures to exercise strict control to avoid blending of this concrete with nonrecyclable concrete.

The authors intend to investigate the system further and promote the complete recycling of concrete.

Acknowledgement

The authors express their gratitude to Mr Norio Yokota, former chairman, Mr Masaru Honda, former member, Dr Akira Ooshio, present chairman, and other members of the Construction Byproduct Utilization Promotion Committee, the Japan Cement Association, and Mr Shigeru Takahashi of the Research Institute of the Japan Cement Association, for their assistance.

References

1.   Tomosawa, F., et al. (1994) Study on Completely Recyclable Concrete, Summaries of Technical Papers of Annual Meeting of Architectural Institute of Japan

2.   Project of General Technical Development of the Ministry of Construction (1994) Technological Development of Control of Occurrence and Recycle of Construction By-Product, Reports of Fiscal Heisei 5

3.   The Ministry of International Trade and Industry : The 1989 Survey of Statistics of an Estimated Amount of Mineral.

4.   Japan Cement Association (1993) Summaries of Consideration Progress of Special Committee for Environment, Technical Symposium of Cement Production, Cement Journal, 3 January 1994

5.   Tomosawa, F. (1994) The Appearance of Completely Recycling Concrete, Nikkei Architecture No.506, 21 November 1994

25 ECOLOGICAL CONCRETE WITH CONTINUOUS VOIDS

M. TAMAI
Kinki University, Higashi-Osaka City, Japan

Abstract
No-Fines concrete (NFC), with continuous voids through which water and/or air can pass freely, is a much more attractive type of concrete and can be rendered for a wider variety of functions than ordinary concrete, which generally requires high mechanical properties. Various attempts have been made to evaluate and expand the uses of NFC. For example, a function of purifying water can be given to NFC by fixedly applying various proper microorganisms to the inner and outer surfaces of the NFC and allowing them to live therein. NFC boards which have noise-absorbing functions of absorbing at particular frequencies are obtained by controlling the thickness of the NFC boards or their porous structures. NFC which has the functions of heat storage and radiation is obtained through the use of heavy aggregates.

NFC, through which water and/or air can pass freely, is assumed to be suitable for organisms. In this report, void ratios, water permeabilities and compressive strengths of NFC, measured when the types of binders including cement and the grains of aggregates were changed, are discussed. Also, attachment of organisms in sea water, fixing and growing of lawns, weeds, and liane on NFC are investigated.

Generally, when such concrete containing ordinary Portland cement as a binder is placed at water's edge, free lime is eluted during the initial test period. This tends to exert a hazardous effect on animals and plants, and reduces its service life.

However, if ordinary cement with silica fume or blast furnace slag cement is used as a binder to constitute porous concrete, their defects are modified. As the results, NFC is thought to be useful in the establishment of well-balanced biological environment.
Keywords: Attachment, continuous voids, marine, no-fines concrete, organism, planting concrete.

Integrated Design and Environmental Issues in Concrete Technology. Edited by K. Sakai. Published in 1996 by E & FN Spon, 2–6 Boundary Row, London, SE1 8HN, UK. ISBN 0 419 22180 8.

1 Introduction

Recently, the atmospheric concentrations of carbon dioxide, nitrogen and sulfur oxides are increasing more than ever due to the large-scale deforestation of tropical rain forests and burning of fossil fuels. These gases are now causing climatic warming of the earth and acid rain, and rapidly deteriorating global environment. In addition, waters of lakes and marshes, rivers and oceans are also being polluted by the wastes from industries and human activities, and their natural self-purification activities are severely damaged by overloading and the destruction of ecosystems.

Cement concrete has been used as a construction material primarily for various kinds of large public buildings, bridges, roads, structures and facilities. It has contributed not only to the enlargement of social infrastructures, but also to cultural and economic development to a great degree. It is impossible, however, to refute the suggestion that concrete buildings and facilities, which are designed to be purely functional and highly convenient, have been responsible for the destruction of the natural environment, namely, greenery and watersides, and threaten the survival of endangered plant and animal species which inhabit those places. For example, the concrete structures used for river revetments, newly-built highway earth-retaining walls and hotel facilities associated with the development of the resort areas may cause those species living nearby to be greatly reduced in number, even to the point of extinction.

Research of No-fines concrete(NFC) is being performed media for to improve the strength and other properties of this type of concrete, and success has achieved in obtaining desired strength and other mechanical properties, including the resistance to freezing and thawing[1][2][3].

The purpose here is to explore the NFC a little further. It is a new type of functional material, so called because it allows bacteria, small animals and plants to live in and grow together in harmony with the natural environment which is to be protected and conserved[4].

This paper gives a thorough description of the properties of some types of NFC, in addition to the manufacturing method, and indicates a few examples of its application.

2 Requirements for constituting of NFC

2.1 Forming conditions for continuous voids

Suppose that the process by which these kinds of constituents are filled, has three separate phases, as shown in Table 1; namely aggregate (solid phase), cement paste (liquid phase) and air (air phase). A basic requirement for producing porous concrete, whose cells may be open and interconnecting, is that the mixture be prepared so that its properties can be found in the Funicular first range (F-1), where all the phases coexist and have a fixed order, especially when each phase is connected to the next one. This applies equally well to the Funicular second range (F-2), where the solid and liquid phases are connected to each other, both are independent of the air phase, and have a fixed order, as long as F-2 covers part of the F-1 above[2].

Table 1. Rheological filling pattern

Filling Pattern	Pendular range	Funicular range		Capillary range	Slurry range
		F-first range	F-second range		
Solid phase	continuous	continuous	continuous	uncontinuous	uncontinuous
Liquid phase	uncontinuous	continuous	continuous	continuous	continuous
Air phase	continuous	continuous	uncontinuous	0	0
Condition	 Paste	 Continuous void	 Independent air	 Paste	 Paste

2.2 Filling models of balls and critical void ratio of F-2 range

A container filled with crushed stone with a high uniformity coefficient usually has a void ratio range of 39~47%. Good approximations can be obtained by considering this stone system as a combination of simple cubic lattices, filled with equal-sized particles (void ratio: 39.54%) and simple stagger lattices, filled with equal-sized particles (void ratio: 47.64%). Suppose that in both models spherical aggregates are covered with a coat of stiff cement paste which does not cause it to segregate. The maximum air bubble diameter, obtained in the case of 40% of the voids being filled with that cement paste, and the corresponding void ratio, obtained in the process of transferring from F-1 to F-2, are shown in Table 2. This table shows that no open-structured cells can be produced unless the material has a void ratio of 19.75% or more for simple cubic lattices filled with equal-sized particles and 8.89% or more for simple stagger lattices filled with equal-sized particles, respectively. When the diameter of the sphere is designated R, the maximum radius of those isolated pores is equal to 0.723R and 0.528R, respectively.

From what has been said above, it may be reasonable to conclude that material made from single-grained aggregates should have a void ratio of 20% or more in order of the voids to be open and interconnecting.

Table 2. Filling model of balls and void ratio

Packing model	Number of ball contact point	Void ratio (%)	Largest radius of independent air	Void ratio of critical point(F-2), (%)
Most coarse	4	66.00	-	-
Simple cubic	6	47.64	0.723R	19.75
Simple stagger	8	39.54	0.528R	8.89
Double stagger	10	30.19	0.471R	7.70
Pyramidal	12	25.95	0.414R	5.25
Most dence	12	25.95	0.225R	1.25

3 NFC manufacturing method and its properties

3.1 Materials used and concrete mixture proportions

Ordinary Portland cement (CN), B-type slag cement (CBb), C-type slag cement (CBc) and fly ash cement (CF) can be used to make the NFC. It is preferable to use either cements which contain less tricalcium silicate (C_3S) or cements mixed with pozzolan. This is due to the fact that neither should elute any free lime, have no effect on living things or cause reductions in the durability of the concrete. It is suitable to use crushed stones No.5 (20~13 mm), No.6 (13~5 mm), No.7 (5~2.5mm), No.8(2.5~1.2mm) and equal-sized crushed stone (G).

What is most important is that the mixture proportions of NFC depend on their strength and the amount of voids that may be open and interconnecting. These voids should be produced in such a way that any plant or greenery could sink its roots into and through them easily. Crushed stone No.5 is recommended as the most suitable for use. In raising small animals and microorganisms, crushed stone No. 6 or a large one is recommended for use as an aggregate in view of its large water permeability.

Typical NFC mixture proportions fit for living things are given in Table 3. It is reasonable to conclude that one of the best mixture proportion methods is to fill an adequate amount of binder into the voids of the aggregate. The neat cement paste should have a filling ratio of 25~50% to the crushed stones voids, as long as it is used for practical purposes.

Table 3. Mixture proportions of NFC

Mixing Type	Kind of binder	SF/(C+SF) (%)	W/(C+SF) (%)	Unit weight (kg/m³)				
				W	C	SF	G	Sp
A	CN	-	25	72.8	292	-	1560	2.8
B	CB	-	25	71.5	286	-	1560	2.1
C	CN+SF	20	25	68.8	220	55	1560	4.6

CN : Ordinary Portland cement(Sg. : 3.16), SF : Silica fume(Sg. : 2.20)
CB : Portland blast-furnace slag cement(Sg. : 3.03)

In all cases where it is necessary to increase the strength of the NFC, it should be produced with water-cement ratio of 20~30% by the addition of an adequate amount of superplasticizer (Sp) at the same time. Furthermore, in order to improve the rheological properties and strength of the cement paste bonded to the aggregate, it is proper to add pulverized pozzolanic filling materials such as silica fumes (SF) and superplasticizers to the mix.

3.2 Manufacturing method

NFC could be produced by mixing all the concrete ingredients at once by means of a forced action mixer. However, the quality of the mixture can be improved by mixing the binder with the cement paste mixer, subsequently determining the required amount of cement and aggregate for each batch, and finally mixing the ingredients all together by a forced action mixer on a separate basis.

The planting beds are made from NFC, whose pores on the surface have been filled in with such good earth as vermiculite, and have a great water holding capacity. Plants can be grown directly on these beds.

3.3 Methods for testing the properties of NFC

(1) Permeability test: conducted with a 30 cm head (H) of in conformity to JIS A 1218;
(2) Compressive strength test: conducted in conformity to JIS methods;
(3) Elution test of free lime: specimens with $\phi = 10 \times 20$ which have undergone a variety of curing and are then kept in 2 liters of clean water over 24 hours for curing; the hydrogen ion concentration can be determined by a pH meter.

3.4 Properties of NFC

3.4.1 Rheological property of fresh cement paste
The point of manufacturing NFC, whose voids may be open and interconnecting, is to obtain a concrete mixture with the proper consistency in order to avoid segregation. This will also depend upon the rheological and filling properties. The consistency does not change even when vibrations are created during manufacturing, and at the same time the formation of the Funicular first range is a requirement. For example, if those rheological properties such as the plastic viscosity and yield value of the cement paste become smaller in value, they drip from the aggregates. If they are too stiff, a uniform film of cement can not form on and bond to the surface of the aggregate.

When crushed stone No.6 (13~5 mm) is used as an aggregate, the yield point of the cement paste shall be more than 20~80 Pa, while the plastic viscosity will be more than 10 Pa•s.

3.4.2 Water permeability
Table 4 shows the relationship between water permeability and the voids of NFC using three types of aggregates. Based on the the results, it is apparent that the NFC made with a filling factor of 50% or less from an aggregate with a grain size of No.7 or larger can be used as a means of raising animals and plants. It is desirable that the decision as to which kind of animal or plant will be able to live and grow be dependent on the size of the NFC voids, rather than on the permeability. Such decision can be made most properly by choosing the size of the NFC voids whenever necessary.

Table 4. Water permeability of NFC(mm/s)

Kind of crushed stone	Voids ratio of NFC(%)		
	30	25	20
No.6	17.5	12.3	6.5
No.7	10.5	7.5	4.0
No.8	4.6	2.7	1.3

3.4.3 Strength properties
Table 5 shows the relationship between the compressive strength and the material age of the NFC made from cements (CN, CB, CN+SF) and crushed stones No.5 and No.6. If cement CN is used, it means that the NFC made from this will show a stable increase

in strength from its early life for a long time to come. On the other hand, if CB and CN+SF are used, this NFC will have less initial strength, but exhibits a sharp increase in long term strength. Furthermore, it is found that the compressive strength varies inversely with the grain size of the aggregate. This phenomenon is probably related to the fact that the number of contacts per unit volume of aggregate sharply decreases. Due to the absence of bleeding of and the formation of a very good bond between the aggregate and the cement, the NFC exhibits a considerably higher values for compressive strength, bending and compressive strength ratio better than ordinary concrete.

Table 5. Compressive strength vs. age

Kind of mixing		Compressive strength (MPa)		
		7 days	28 days	56 days
A	*	10.7	12.7	13.5
B	No.5	8.6	11.4	12.7
C		9.1	12.5	14.8
A	*	14.8	18.6	20.5
B	No.6	12.2	17.5	20.1
C		13.5	20.2	21.8
A	*	16.3	19.7	21.6
B	No.7	14.4	18.2	21.4
C		15.2	20.6	23.3

* : Kind of crushed stone

3.4.4 Free lime elution

Storm water or immersed in water often causes free lime to be eluted intensively from NFC because the internal surface area of this concrete, as well as its roughness, is very large. Such water may sometimes have a bad influence on animals and plants, especially when they are young. It also causes the thin film of cement, put on the surface of the NFC, to sharply deteriorate. Thus, its useful life is shortened. In view of these facts, it is desirable that the drawback to using this concrete should be improved by the addition of silica fume to the cement, by use of slag cement and by the reaction of pozzolan to form the concrete, so as to prevent the elution of free lime. Table 6 shows experimental results of the NFC specimens made from crushed stone with three different mixture proportions and cured river water for periods of 14~180 days, and measured pH.

Table 6. pH of NFC immersed in river

Kind of Mixing		pH				
		14 days	28 days	56 days	91 days	180 days
A	*	10.6	9.9	9.3	8.8	8.5
B	No.6	10.4	9.6	9.1	8.6	8.2
C		10.0	9.3	8.8	8.5	8.2

* : Size of crushed stone(13~5mm)

4 Attachment of marine organisms

4.1 Cycle of self-purification process performed by marine organisms

As shown in Fig.1, the organic matters produced in sea water are decomposed and mineralized into CO_2, H_2O, NH_4^+ etc. by aerobical heterotrophic bacteria. Ammonia is highly toxic to fish and other marine lives, but, when enough amount of dissolved oxygen is available, it is oxidized to nitric acid by nitrifying bacteria. Algae then use this inorganic nitrogen and other nutritional minerals for photosynthesis, reproducing organic matters by a process called primary production.

The plants such as algae produced by this process are then ingested by marine animals and moves the cycle one step forward to the higher production. When oxygen availability is limited, however, sulfate in the sea water is reduced to hydrogen sulfide by the activity of sulfate-reducing bacteria. This activity consumes dissolved oxygen and therefore results in the deterioration of marine environment. Thus, the activities of aerobical heterotrophic bacteria and nitrifying bacteria are desirable for the substrate cycle in such sea water area, and, therefore, development of environment-stimulating materials that enhance the growth of these bacteria is necessary.

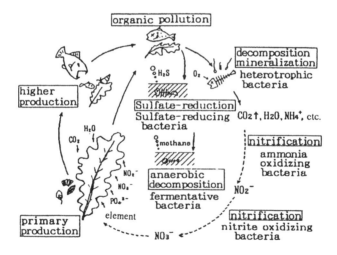

Fig.1 Cycle of self-purification process in sea water

4.2 Attachment of organisms

When the concrete test pieces (ϕ 15 x 30 cm) compose of the materials presented in Table 4 were sunken to the depth of -100 to -150 cm from mean sea water level in the natural sea along the shore, various kinds of marine organisms attached and proliferated on the interior and exterior surface of NFC. Thus, a good biological environment was established in a short period.

(1) Bacteria
Neutralization of the NFC surface progressed in 10 days after sinking into sea water and

microbial adhesion was started. It is assumed that it takes about one month to neutralize the internal cavities by inflowing of sea water. Attachment and proliferation of aerobical heterotrophic bacteria began around the time of completion of neutralization and slightly later, nutrifying bacteria developed and proliferated on them. Thus, many kinds of organic substances in the sea water were actively metabolized by those bacteria.

(2) Unicellular algae

Various kinds of unicellular algae attached on the surface exposed to sunlight shortly after the attachment of bacteria. Most of the algae on the surface were found to be diatoms. From the time of this stage, the amount of bacteria on it began to decrease gradually.

(3) Multi-cellular algae

Attachment of Ulva pertusa started around the 14th-day after sinking of concrete specimen. Attachment of Gracilaria compressa and Codium fragile were occasionally noted in some area after nearly 3 months.

(4) Small marine animals

Attachment and invasion of ciliates, nematodes, polychaetes and crustaceans were found after 20 days or so and in early time, larvae of crustaceans adhered and invade the column preferentially. After two months or so, it was observed that a large number of nereis invaded to the internal surface.

(5) Shellfishes

Attachment of Crassostrea gigas and Chthamalus challengeri began nearly 30 days after immersion, followed by attachment of Soletellina diphos. In general, adhesion behaviors of marine organisms were greatly varied depending on seasonal factors. The layers made of these organisms were further multiplied by interacting each other after 3~6 months, resulting in different growing conditions depending on the time of its sinking. After one year or more, a new environment similar or rather superior to the biological environment in the surrounding sea area was established.

Photo 1 shows the test specimen (ϕ 15 x 30 cm) composed of the NFC before immersion and Photo 2~6 shows the states of the ordinary and NFC specimen immersed in the sea for a period of 30 days to 1 year.

Photo. 1 NFC specimen before immersion   Photo.2 NFC specimen 30 days immersion

Photo. 3 Normal concrete
(30 days immersion)

Photo. 4 Attachment of <u>Crassostrea gigas</u>
and <u>Chthamalus challengeri</u>

Photo. 5 Attachment of sponge and <u>Ulva pertusa</u>
(1 year immersion)

Photo. 6 Attachment of <u>Codium fragile</u>
(1 year immersion)

5 Inhabitability of plants

(1) Lawn grass vegetation
River sand was spread to a depth of 1 cm on the surface of the NFC specimen (100 x 100 x 10 cm). After block sodding with Zoysia matrella, the growing conditions were evaluated. The block sodding on the NFC was found to be favorable for producing a good biological environment because of the large drainage capacity of the NFC. After 3 months, it was found that the roots of lawn grasses had penetrated the continuous porous cavities of the NFC.

(2) Weeds
After transplantation and the cuttage and seeding of various kinds of weeds, the germination and the growing conditions were examined for the surface of NFC specimen (30 x 30 x 10 cm).

As the subjects for transplantation and cuttage, Plantago and Tradescantia flumiensis were used, respectively. For the germination test, Plantago and Anthraxon were used as the subjects.

The results show that when crushed stone No.5 was used as one of the compound components, the transplantation of Plantago led to the most satisfactory results and the planted cuttings of Tradescantia flumiensis also grow well on the surface. Moreover, the root spreadings and germination rates of Plantago and Anthraxon were examined on the concrete containing the crushed stone No.5. And several types of weeds such as Portulaca were found to have grown naturally in the area surrounding the NFC specimen.

Photo.7 Condition of sodded a lawn in
NFC after 3 months

Photo.8 Condition of transplanted
Plantago in NFC after16 days

Photo.9 Condition of planted
Tradescantia cutting after 26 days

Photo.10 Condition of seeded
Arthraxon hispidus

6 Conclusions

Since it is possible for air and/or water to pass through porous concrete of a cellular structure, with numerous continuous cavities whose size and strength can be controlled, NFC is thought to be utilizable as a habitat for plant and animal organisms. In particular, when NFC was sunken along the shores of sea, a river and a lake, microorganisms and small animals as well as plants could inhabit it. Accordingly, NFC is expected to generate a good environment for living organisms.

7 References

1. Malhotra, V.M. (1976) No-Fines Concrete - Its Properties and Applications, Jour. of ACI, Vol.73, No.11, pp.628-644
2. Tamai, M. (1988) Water Permeability of Hardened Materials with Continuous Voids, CAJ Review, pp.446-449
3. Tamai, M. (1989) Properties of No-Fines Concrete Containing Silica Fume, ACI, SP-114, Vol.2, pp.799-814
4. Tamai, M. and Nishiwaki, Y. (1992) Studies on Marine Epilithic Organisms to No-Fines Concrete Using Slag Cement and Portland Cement with Silica Fume, ACI, SP132-87, Vol.2, pp.1621-1635

26 HIGH-PERFORMANCE CONCRETE INCORPORATING LARGE VOLUMES OF LOW-CALCIUM FLY ASH

V.M. MALHOTRA
CANMET, Natural Resources Canada, Ottawa, Canada

Abstract
This paper outlines the background to the development of high-performance concretes incorporating high volumes of low-calcium (ASTM Class F) fly ashes. Briefly, this type of concrete incorporates about 55 to 60 per cent ASTM Class F fly ash as replacement for portland cement, and low water-to-cementitious materials ratio of about 0.30 is achieved by the use of large dosages of superplasticizers. This paper provides data on the mechanical properties and durability aspects of high-volume fly ash concretes. The mechanical properties investigated include compressive and flexural strengths, modulus of elasticity and creep; the durability aspects covered include resistance to freezing and thawing, and resistance to water and chloride-ion penetration. The test data show that high-volume fly ash concrete with very low portland cement content is a high-performance system, and its mechanical and durability properties are equal or superior to that of the control portland-cement concrete with significantly higher amounts of portland cement.
Keywords: Concrete, compressive strength, durability, fly ash, high performance, permeability.

1   Introduction

Canada Centre for Mineral and Energy Technology (CANMET) has an on-going project in the area of supplementary cementing materials for use in concrete. The major objective of this project is to conserve energy, and to develop environmentally friendly concretes incorporating industrial by-products such as fly ash, slag and silica fume. As a part of this project, CANMET has developed high-performance concrete incorporating large volumes of low-calcium ASTM Class F fly ash. Briefly, this

Integrated Design and Environmental Issues in Concrete Technology. Edited by K. Sakai. Published in 1996 by E & FN Spon, 2–6 Boundary Row, London, SE1 8HN, UK. ISBN 0 419 22180 8.

concrete has water-to-cementitious materials ratio of about 0.30, and cement and fly ash contents of about 150 and 215 kg/m^3, respectively. The slumps >150 mm are achieved by the use of large dosages of a superplasticizer. The concretes, so developed, have all the attributes of high-performance concrete i.e. excellent long-term durability characteristics, low permeability, and high strength (1-5).

This paper gives typical mixture proportions of this concrete, and discusses its mechanical properties and durability aspects.

2   Properties of cement, fly ash and superplasticizers

A variety of ASTM Type I and Type III cements, ASTM Class F fly ashes and naphthalene-based superplasticizers have been used in this project.   Some typical physical properties and chemical analysis of the cements and fly ashes used are shown in Tables 1 and 2.

Table 1. Physical properties and chemical analysis of cement[*]

	ASTM Type I Cement	ASTM Type III Cement
Physical Tests		
Fineness - passing 45 μm, %	94.9	97.6
- Blaine, m^2/kg	376	608
Specific Gravity	3.14	--
Compressive strength of		
51-mm Cubes, MPa: 3-day	31.5	37.1
7-day	34.5	--
28-day	41.9	48.0
Chemical Analysis		
Silicon dioxide (SiO$_2$) %	19.20	20.69
Aluminum oxide (Al$_2$O$_3$) %	5.79	4.87
Ferric oxide (Fe$_2$O$_3$) %	2.03	2.01
Calcium oxide (CaO), total %	63.48	63.25
Magnesium oxide (MgO) %	2.52	3.22
Sulphur trioxide (SO$_3$) %	3.50	3.96
Sodium oxide (Na$_2$O) %	0.33	--
Potassium oxide (K$_2$O) %	1.16	--
Loss on ignition	2.61	1.38
Bogue Potential Compounds		
C$_3$S	63.7	53.3
C$_2$S	7.0	19.2
C$_3$A	11.9	9.5
C$_4$AF	6.2	6.1

[*]From references 2 and 5.

Table 2. Range in physical properties and chemical analysis of some of the fly ashes used in the investigation[*]

	Range
Physical Tests	
Fineness - passing 45 μm, %	68.1-85.6
- Blaine, m^2/kg	221-348
Specific Gravity	2.04-2.63
Chemical Analysis	
Silicon dioxide (SiO$_2$) %	42.20-55.39
Aluminum oxide (Al$_2$O$_3$) %	15.6-31.53
Ferric oxide (Fe$_2$O$_3$) %	3.52-27.60
Calcium oxide (CaO), total %	1.17-14.93
Magnesium oxide (MgO) %	0.85-4.34
Sulphur trioxide (SO$_3$) %	0.09-1.85
Sodium oxide (Na$_2$O) %	0.21-5.52
Potassium oxide (K$_2$O) %	0.50-2.55
Loss on ignition	0.57-3.70
Carbon, %	0.28-1.65
Activity with Cement, %	86-97

[*]From reference 2.

The superplasticizers used were sulphonated naphthalene-formaldehyde condensates of Japanese and Canadian origin, with about 40 per cent solids and densities of about 1200 kg/m^3. Several commercially available air-entraining admixtures were incorporated in the concrete mixtures.

3   Mixture proportions

The basic principle of the mixture proportions was to produce high-performance, air-entrained concrete for structural applications. Typical mixture proportions are shown below.

Typical Mixture Proportions	
ASTM Type I or Type III Cement	$= 150$ kg/m^3
ASTM Class F Fly Ash	$= 210$ kg/m^3
Water	$= 120$ kg/m^3
Coarse Aggregate	$= 1195$ kg/m^3
Fine Aggregate	$= 645$ kg/m^3
Air-entraining Admixture	$= 210$ mL/m^3
Superplasticizer	$= 4.5$ L/m^3

4 Properties of fresh and hardened concrete

4.1 Fresh concrete

In this type of concrete, the slumps are adjusted by varying the dosage of the superplasticizer. Most of the investigations at CANMET have been done with "flow" slumps i.e. slumps between 200 and 250 mm. The entrained-air contents are kept between 5 and 7 per cent. It has been found that this percentage of air usually results in satisfactory bubble-spacing factor, $\bar{L}$, in the hardened concrete, i.e. $\bar{L}$ is less than 250 μm. The setting time of the high-volume fly ash concrete is somewhat longer than that of the control concrete but this is of little practical consequence.

4.2 Hardened concrete

The typical properties of the hardened concrete are shown below.

Typical Properties of Air-Entrained Hardened Concrete		ASTM Type I Cement	ASTM Type III Cement
Compressive Strength:	1-day	$= 8\pm2$ MPa	14 ± 2 MPa
	28-day	$= 35\pm5$ MPa	40 ± 4 MPa
	91-day	$= 45\pm5$ MPa	50 ± 5 MPa
	365-day	$= 55\pm5$ MPa	--
Flexural Strength:	14-day	$= 4\pm0.5$ MPa	5.5 ± 0.5 MPa
	91-day	$= 6\pm0.5$ MPa	6.5 ± 0.5 MPa
Splitting-Tensile Strength:	28-day	$= 3\pm0.5$ MPa	3.5 ± 0.5 MPa
"E" Modulus at 91 Days		$= 35\pm2$ GPa	37 ± 2 GPa
Drying Shrinkage Strain at One Year		$= 500\pm50\times10^{-6}$	--
Specific Creep Strain at 365 Days		$= (28\pm3)^x10^{-6}$	--

Figure 1 shows a typical compressive strength versus age relationship for high-volume fly ash concrete for ASTM Type I cement. Considering that the concrete contains only 150 kg/m³ of cement, the later-age strengths at 90 days and one year are very high. The Young's modulus of elasticity "E" is of the order of 35 GPa at 90 days. This high modulus is achieved due to the fact that a considerable portion of the unreacted fly ash, consisting of glassy spherical particles, acts as a fine aggregate (3).

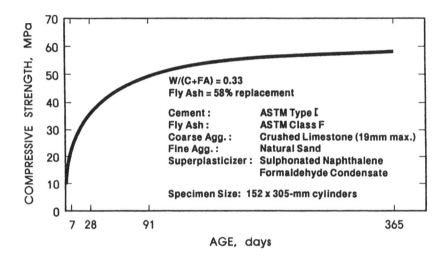

Fig. 1. Typical compressive strength development of concrete made with ASTM Type I cement and incorporating ASTM class F fly ash. (From reference 2).

5   Durability aspects

This new high-performance concrete has excellent durability characteristics. Some of these are discussed below:

5.1 Resistance to the penetration of chloride ions
The new concrete being discussed has very high resistance to the penetration of chloride ions as determined by ASTM C 1201 test method. The charge measured in coulombs is of the order of 500 at 90 days. Normally, in this test which had been developed to measure indirectly the permeability of silica fume concretes, a value of less than 600 coulombs is indicative of very low permeability. The difference between the permeability of silica-fume and high-volume fly ash concretes is that the low permeability is achieved at early ages in the former concrete than in the latter concrete (1,4).

5.2 Water permeability
The water permeability of the high-performance fly ash concrete is very low. Tests performed under uniaxial flow conditions under a pressure of 2.7 MPa have indicated

the permeability to be of the order of 10^{-13} m/sec (1). The tests are performed on 50-mm thick concrete disks using a CANMET developed uniaxial flow test. The low permeability of the concrete is due to the very low porosity of the portland cement/fly ash concrete system with gel pores being discontinuous.

5.3 Resistance to freezing and thawing
The air-entrained, high-performance fly ash concrete shows excellent resistance to repeated cycles of freezing and thawing. Even after 1000 cycles in ASTM C 666 "Procedure A" test methods (freezing and thawing in water), the durability factors are in excess of 90 (2). Apart from some surface scaling, the test prisms show no distress of any kind. Normal portland cement concrete are considered satisfactory if they can survive 300 cycles of freezing and thawing in the above test.

5.4 Thermal properties
Because of the very low cement content, the temperature rise in the high-performance, high-volume fly ash concrete is very low. In one investigation several large monoliths, 2.5x4.0x5.0 metre in size, were cast in Laval, Quebec to monitor the temperature rise of concrete in the interior of the monoliths. A maximum temperature of 54°C was recorded at the mid-height of the monoliths after 7 days of placement; the ambient temperature and the temperature of concrete at the time of placing were 22 and 21°C, respectively (1). Thus, the high-volume fly ash concrete is ideally suited for structural concrete, and for mass concrete such as dams, mat foundations and large retaining walls.

5.5 Control of expansion due to alkali-aggregate reaction
The undesirable expansion of concrete due to reaction between the cement alkalies and certain types of silica in aggregates is a serious problem. Research at CANMET and elsewhere has shown that the above alkali-silica reactions (ASR) in concrete can be controlled by incorporating good quality fly ash as a partial replacement for cement. The general recommended levels of cement replacement by fly ash are between 25 and 40 per cent. Extensive tests performed at CANMET have shown the use of high-performance, high-volume fly ash concrete can reduce effectively expansion due to ASR. This has been shown to be so by both standard CSA and the accelerated test methods. The rationale for reduction in the expansion due to ASR is the extremely low permeability of the high-volume fly ash concrete system, and the reduced pH value of the pore solution.

5.6 Carbonation
Carbonation of high-volume fly ash concrete is not an issue because of its very low water-to-cementitious ratio which results in very dense matrix. Limited tests performed on the broken portions of the cores drilled after eight years from a large block of high-volume fly ash concrete have shown the carbonation depth of about 10 mm.

6 Limitations of high-performance, high-volume fly ash concrete

Notwithstanding the excellent mechanical properties and durability characteristics of the high-volume fly ash concrete, there are certain precautions one must take when using this concrete. These include control of the quality of fly ash, and the compatibility of the portland cement/fly ash/superplasticizer system. Also, the adequate curing of the concrete is most essential otherwise full strength potential and low permeability of the concrete may not be achieved.

7 Concluding remarks

The high-performance, high-volume fly ash concrete described in this paper has good potential for use in structural and mass concrete applications. Adequate curing is essential if the concrete is to achieve its full potential.

8 References

1. Malhotra, V.M. (1992) CANMET Investigations Dealing with High-Volume Fly Ash Concrete; published in *Advances in Concrete Technology*, CANMET Report 92-6 (R), Energy, Mines and Resources Canada, Ottawa, pp. 433-470. (Available from: CANMET/EMR, Ottawa, Canada).
2. Carette, G.G., Bilodeau, A., Chevrier, R., Malhotra, V.M. (1992) Mechanical Properties of Concrete Incorporating High-Volumes of Fly Ash from Sources in the U.S.A., Electric Power Research Institute (EPRI), Report TR 100577, Project 3176-6 *Proceedings*, Palo Alto, CA 94304.
3. Sivasundaram, V., Carette, G.G. and Malhotra, V.M. (1989) Long-Term Strength Development of High-Volume Fly Ash Concrete, *MSL Division Report MSL 89-53 (OP)*, Energy, Mines and Resources Canada, Ottawa.
4. Malhotra, V.M., Carette, G.G., Bilodeau, A.A. and Sivasundaram, V. (1991) Some Aspects of Durability of High-Volume ASTM Class F (Low-Calcium) Fly Ash Concrete, *ACI Special Publication SP 126*, pp. 65-82. (Ed. V.M. Malhotra).
5. Bilodeau, A. and Malhotra, V.M. (1993) High-Performance Concrete Incorporating Large Volumes of ASTM Class F Fly Ash, *Division Report MSL 93-60 (OP&J) Draft*.

27 CHEMICAL ADMIXTURES FOR HIGHLY FLOWABLE CONCRETES

T. KAWAI
Institute of Technology, Shimizu Corporation, Tokyo, Japan

Abstract
Firstly, this paper introduces chemical admixtures including newly developed one for highly-flowable concretes. These chemical admixtures are categorized "high-range water-reducing agent", "superplasticizer" and "air-entraining and high-range water-reducing agent". These chemical admixtures contribute greatly to the realization of recent epoch-making technologies called "highly-flowable concretes".

Secondly, main features of highly-flowable concretes are described. Highly-flowable concretes assure to enhance the performance and to improve the constructability.

Finally, the contributions of highly flowable concretes to environmental issues are presented.

Keywords: Air-entraining and high-range water-reducing agent, antiwashout underwater concrete, chemical admixtures, high-range water-reducing agent, high-strength concrete, highly flowable concrete, superplasticizer.

1 Introduction

V.M.Malhotra[1] traced development of high-range water-reducing agent in Japan and that of superplasticizers in Germany in a review paper. In Japan, Hattori and co-workers[2] pioneered the development of formaldehyde condensates of beta-naphthalene sulfonates type "high-range water-reducing agent" with the primary aim of significantly reducing the water demand of concrete in order to produce "high-strength concrete". Water reductions of up to 30 percent were able to be achieved with the use of the agent. In Germany, Aignesberger and his colleagues [3] developed the melamine based superplasticizer with the primary aim of producing "flowing concrete". Thereafter, melamine type high-range water-reducing agent and naphthalene type superplasticizers were also developed.

Integrated Design and Environmental Issues in Concrete Technology. Edited by K. Sakai. Published in 1996 by E & FN Spon, 2–6 Boundary Row, London, SE1 8HN, UK. ISBN 0 419 22180 8.

Since then, the two chemical admixtures have been used worldwide. The aims of these two chemical admixtures are different from one another. The ingredients are however nearly the same. International conferences on superplasticizer including high-range water-reducing agent organized by CANMET and ACI have been held four times. And a fifth one is planned for 1998.

Furthermore, the above two chemical admixtures have been standardized in different countries. Quality specifications are prescribed by such as ASTM C 494-92 (Chemical Admixtures for concrete, Type F and Type G), ASTM C 1017-92 (Chemical Admixtures for Use in Producing F.C. Type 1 and Type 2), BS 5057 : Part 3 :1985 (Specification for Superplasticizing Admixtures, Mix A and Mix B) and CAN3-A266.6-M-85(Superplasticizing Admixtures for Concrete SPN TYPE and SPR TYPE).

These two chemical admixtures however have one common defect that the slump loss is considerable. To solve this problem, new technology of slump control with a reactive polymeric dispersant[4] and steric hinderance theory [5] were studied. As a result, "air-entraining and high-range water-reducing agent" was also developed in Japan. This agent can be mixed into the mixture in the ready-mixed batching plant and the concrete can be transported to the job-site to be placed because the slump loss is relatively small. Since its introduction, the use of air-entraining and high-range water-reducing agent has increased dramatically in Japan. Under the circumstances, in 1995 air-entraining and high-range water-reducing agent was added to be prescribed into JIS A 6204 (Chemical Admixtures for Concretes). Its quality specifications are shown in Table 1.

Table 1. Quality specifications for air-entraining and high-range water-reducing agent (JIS A 6204)

Item	Standard type	Retarding type
• Reduced water content min. % of control	18	18
• Bleeding max. % of control	60	70
• Time of setting, allowable deviation from control, min.		
Initial: at least	30 earlier	90 later
not more than	nor 120 later	nor 240 later
Final: at least	30 earlier	
not more than	nor 120 later	120 later
• Compressive strength, min. % of control		
3 days	135	135
7 days	125	125
28 days	115	115
• Length change, max.	110	110
• Relative durability factor, min.	80	80
• Slump change in 60 min., max., cm	6	6
• Air content change in 60 min., max. %	±1.5	±1.5

Here, I will summarize principal functions of the chemical admixtures. They are as follows:

Reducing water cement ratio to get high strength.
Reducing water content to improve durability.
Reducing cement content to prevent thermal cracks.
Increasing fluidity to improve constructability.

These functions are utilized to get highly-flowable concretes.   Next, I will describe the main features of the highly-flowable concretes.

2   Main features of newly developed highly-flowable concretes

Recently various new technologies of concrete construction and of quality improvement have been developed to the stage that they are now widely used.   These new technologies mean high-strength concrete, high-performance concrete, highly-durable concrete, highly-flowable concrete and antiwashout underwater concrete and so on. The definitions of these kinds of concretes are sometimes vague or sometimes different between countries.   For example, high-performance-concrete   was introduced by the ACI paper [6].   But in Japan high-performance-concrete   has been defined by Prof. Okamura as the concrete that needs no vibrating compaction, and has good short and long-term durability.

Here, highly-flowable concrete will be clearly defined in this paper.   Referring to the report of JCI committee [7], in this paper highly-flowable concrete is defined as the concrete that has both high fluidity and adequate segregation resistance   at the same time whether it is placed in air or into water.   Highly-flowable concretes are classified as shown in Fig. 1.   The concretes commonly have high fluidity with more than slumpflow of 50cm.

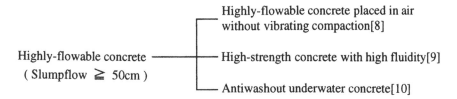

Fig. 1   Classification of highly-flowable concrete.

The most important item I would like to emphasize is that these highly-flowable concretes always use one of the chemical admixtures above-mentioned. In other words, highly-flowable concretes have not been able to be realized without the chemical admixtures for concrete.

Hereafter,   principal features of the mix proportions and characteristics of each highly-flowable concrete will be presented.

2.1   Highly-flowable concrete placed in air without vibrating compaction

Table 2 shows one of the typical mix proportions of highly-flowable concrete for mass concrete.   The cement is composed of moderate heat of hydration cement, blast-furnace slag and fly ash. $150kg/m^3$ of lime powder and $7.8kg/m^3$ of chemical admixture are added.   It is one of the main features of highly flowable concrete to use considerably large amount of powders and chemical admixture to get required segregation resistance and flowablility.

Photographs 1, 2 and 3 show the construction of highly flowable concrete at a job-site. Photograph 1 shows the very densely arranged reinforcing cage. Photograph 2 shows that the concrete is placed through a hopper and a hose by its own weight without vibrating compaction.   Photograph 3 shows that only a worker has to move a tip of hose to required spots to get excellent filling of the concrete.   This example means that highly-flowable concrete can reduce manpower.

Table 2   One of the typical mix proportions of highly-flowable concrete

| MS mm | Unit weight,   kg/m³ | | | | | | |
	W	C MC SL FA	LP	S	G	Ad
40	145	260	150	609	1121	7.8

MC : Moderate heat of hydration cement
SL : Blast-furnace slag
FA : Fly ash
Ad : Air-entraining and high-range water-reducing agent

Photograph 1   Densely arranged reinforcing cage

Photograph 2   Concrete placing apparatus

Photograph 3   Concrete placing

2.2   High-strength concrete with high fluidity

Table 3 shows one of the typical mix proportions of high-strength concrete with high fluidity.   9.5-17kg/m^3 of high-range water-reducing agent was added to get the apporoximately 10 cm of slump and 4.0-6.3kg/m^3 of air-entraining and high-range water-reducing agent was added to get 50-60cm of slumpflow.

Strength characteristics and various durability test results will be shown.   Figure 2 shows the relationship between the replacement ratio of the cement by silica fume and the compressive strength of standard specimens at 7 to 56 days.   the effect on the strength of the replacement of silica fume is negligible at the age of 7 days.

However, as the replacement ratio of silica fume increases, the compressive strength increases at the age of 28 and 56 days. The effect on the strength of a replacement ratio of silica fume up to 10% was outstanding. The compressive strength of the concrete with a 10% replacement ratio of silica fume and a 25% of water-to-cementitious material ratio exceeded 100 MPa at the age of 28 days.

Figure 3 shows the relationship between the compressive strength and the modulus of elasticity. In the range of normal strength concrete, the relation approaches that of the formula of ACI 318M-83. In the range of high-strength concrete, the relation is appreciably close to that of the formula of ACI 363R-84, the "State-of-the Art Report on High-Strength Concrete", which indicates that the modulus of elasticity is approximately 40×10^3 MPa at the strength of 100 MPa. The shape of the ascending part of the stress-strain curve up to failure was more linear and steeper for high-strength concrete.

Figure 4 shows the relationship between the drying shrinkage strain and the duration of drying. The final shrinkage strain, which is the strain when the duration of drying is assumed to be infinite, exhibits $540\text{-}610 \times 10^{-6}$ for high-strength concretes and approximately 930×10^{-6} for normal concrete. The drying shrinkage strains of high-strength concrete are much smaller due to the lower water content.

Figure 5 shows the relationship between the creep coefficient and the duration of loading. The coefficients of high-strength concretes at the age of six months are approximately 0.4, which is approximately one sixth of that of normal strength concrete Even considering the difference in stresses exerted, the creep of high-strength concrete is significantly small.

Figure 6 shows the relationship between the relative dynamic modulus of elasticity and the freezing and thawing cycles. The tests were conducted in accordance with ASTM C 666 Procedure A. All of the high-strength concretes used in this study had dynamic moduli of elasticity of 96-99% at 300 cycles and exhibited excellent resistance to freezing and thawing, even if they were non-air-entrained. We attribute this to the greatly reduced freezable water content and the increased tensile strength of high-strength concrete.

Accelerated carbonation tests were conducted for 40 accumulative hours at a concentration of CO_2 gas of 100% and a pressure of 400kPa. The carbonation depth of the normal strength concrete increased in proportion to the accumulative time and reached approximately 7.5 mm for 40 hours. On the other hand, the carbonation depth of each mixture of high-strength concrete showed zero even for 40 hours. The duration of the accelerated carbonation tests for the speciman converts into approximately 60 years under normal atmospheric condition. Thus, judging from the results, it is thought that the carbonation depth of high-strength concrete will be close to zero even after 60 years under normal atmospheric conditions. The fact that the carbonation depth of high-strength concrete is drastically reduced is due to its low porosity and low permeability.

Table 3   One of the typical mix proportions of high-strength concrete

MS mm	W/(C+SF) %	Unit weight, kg/m³				
		W	C	SF	Ad₁	Ad₂
	22	126	517	58	17	6.3
20	25	135	486	54	12	5.0
	28	140	450	50	9.5	4.0

SF : Silica fume
Ad₁ : High-range water-reducing agent
Ad₂ : Air-entraining and high-range water-reducing agent

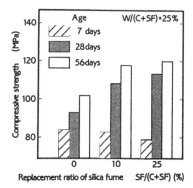

Fig. 2   Compressive strength

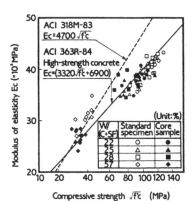

Fig. 3   Modulus of elasticity

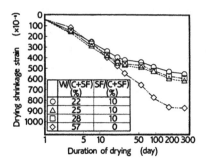

Fig. 4   Drying shrinkage

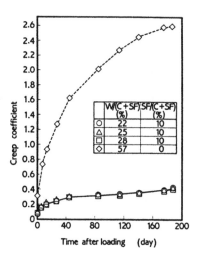

Fig. 5   Creep coefficient

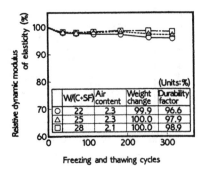

Fig. 6   Freeze-thaw resistance

2.3   Antiwashout underwater concrete

Table 4 shows one of the typical mix proportions of antiwashout underwater concrete. The cement was blast-furnace slag cement Type B (approximately 41% pulverized slag), and the antiwashout admixture was of the cellulose type.   A melamine type high-range water-reducing agent was added.   The target flow property of this concrete was a 50-60 cm slumpflow.

A form, 15 m in length and with transparent sides as shown in Fig.7, was filled with water and the antiwashout underwater concrete was placed from the side to a depth of 1.5 m using a tremie.   The tip of the tremie was equipped with a compressed-air operated valve, by opening and closing which the placing speed was adjusted.   The placing speed, V, of the concrete was made V= 0.1 m/h and V= 0.4 m/h to reflect actual construction data. After every one tenth of the total volume had been placed, the flow gradient of the concrete was measured.

After removal of the formwork , 10cm-diameter cores were taken at 1 m centers and the compressive strength, density, and air content were measured at the age of 28 days. Photograph 4 shows the measurements of flow gradient for a placing speed of V = 0.4 m/h. It can be seen that the flow gradient was large up to about the first quarter of the whole placed volume.   Thereafter, the flow gradient became smaller as the placing height increased..   The reason for the greater flow gradient when the placing volume is small is that the pressure gradient and concrete flow velocity are not proportional.   The final flow gradient after completion of placing remained within 1/500 regardless of the placing speed.   This clearly demonstrates that antiwashout underwater concrete has a self-leveling property.   Figures 8 and 9 show the relationship between underwater flow distance and compressive strength of core specimens and the underwater flow distance and density, respectively.   In Fig. 8 almost no variation in compressive strength with underwater flow distance can be distinguished at V = 0.1 m/h.   On the other hand, it can be seen that the compressive strength decreased somewhat as the underwater flow distance increased at V = 0.4 m/h, though the degree of this decrease was small as long as the underwater flow distance was within 10 m.   In Fig. 9 the density was reduced as the underwater flow distance exceeded 10 m.   The reason for this may be that the volume of coarse aggregate on the surface of the concrete decreases as the underwater flow distance increases.   It was also found that the air content in the core specimens decreased as underwater flow distance increased.   This is explained by the visual observation of air bubbles about 1-5 mm in diameter escaping from the surface of the flowing concrete.   These results demonstrate that antiwashout underwater concrete has a self-compacting property.

These experiments verify that antiwashout underwater concrete with a slumpflow of 50-60 cm has self-leveling and self-compacting properties and that no detrimental segregation was seen as long as the underwater flow distance remained within 10 m.

Table 4   One of the typical mix proportions of antiwashout underwater concrete

MS mm	W/C %	Unit weight, kg/m³					
		W	C	S	G	Aw ad.	Ad.
20	60	220	367	652	1008	2.3	4.6

   C   : Blast-furnace slag cement Type B
Awad : Antiwashout admixture
   Ad   : High-range water-reducing agent

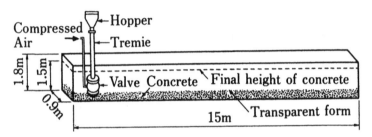

Fig. 7   The form for the long distance flow experiment

Photograph 4   Measurements of flow gradient
for a placing speed of 0.4 m / h

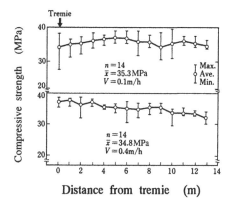

Fig. 8   The relationship between flow distance and compressive strength

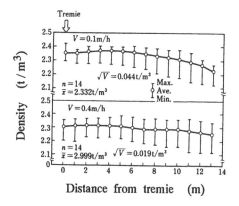

Fig. 9   The relationship between flow distance and density

3   Contribution to global environmental issues

Considering the mix proportions and the construction technology of three highly flowable concretes, these concretes contribute to the global environmental issues in various respects.   The main contributory points are as follows:

(1) Highly-flowable concrete placed in air without vibrating compaction.
- Reduction of cement content reduces production of CO_2.
- Use of mineral admixtures such as blast-furnace slag and fly ash means utilizing by-products.
- Use of mineral admixtures sometimes controls durability such as alkali-aggregate reaction.

(2) High-strength concrete with high fluidity.
· Low water-cement ratio increases the service life of the reinforced concrete structures depending on corrosion of steel due to carbonation and salt attack.
· Low permeability prevents carbonation and ingress of salt.
· Very small amount of entrained air assures high frost resistance.
(3) Antiwashout underwater concrete.
· Almost no segregation during placement undrewater shows no water pollution.
· No water pollution during concreting means protection of environment.
· Construction period is shortened compared with preplaced aggregate concrete.

From the above, it is concluded that highly-flowable concretes and chemical admixtures for them are able to contribute to environmental issues.

4 References

1. Malhotra, V. M, (1989) Superplasticizers:A Global Review with Emphasis on Durability and Innovative Concretes, *Proceedings of 3rd International Conference on Superplasticizers and Other Chemical Admixtures in Concrete,* Ottawa, ACI SP-119, pp.1-17
2. Hattori. K,(1979) Experiences with Mighty Superplasticizers in Japan, ACI *Special Publication* SP-62, pp.37-66.
3. Aignesberger,A and Kern,A.(1981) Use of Melamine-Based Superplasticizer as a Water Reducer", *ACI Special Publication* SP-68, pp.61-80.
4. Izumi.T, Mizunuma.T, Iizuka.M, and Fukuda.M(1989) Slump Control With Reactive Polymeric Dispersant, *Proceedings of 3rd International Conference on Superplasticizers and Other Chemical Admixtures in Concrete,* Ottawa, ACI SP-119, pp.243-264.
5. Nagataki,S.(1990) State-of-the Art on Air-Entraining and High-Range Water-Reducing Agent, JCI Concrete Journal, Vol.28, No.6, pp.5-15
6. Carino,N.J. and Clifton, R.(1991) High-performance concrete: research needs to enhance its use. *Concrete International,* Vol.13, No.9, pp. 70-76
7. Japan Concrete Institute.(1994) *Report No.2 of Research Committee on Highly-Flowable Concrete.*
8. Kawai,T and Okada,T(1989) Effect of Superplasticizer and Viscosity-Increasing Admixture On Properties of Lightweight Aggregate Concrete. *Proceedings of 3rd International Conference on Superplasticizers and Other Chemical Admixtures in Concrete,* Ottawa, ACI SP-119, pp.583-604.
9. Kawai,T et al (1989) Study on Application of 100MPa Strength Concrete Based on Full-Scale Model Tests, *Proceedings of 10th Annual Convention,* SEAOH, Hawaii.
10. Kawai,T(1987) Non-Dispersible Underwater Concrete Using Polymers, *Proceedings of 5th International Conference on Polymers In Concrete,* Brighton, pp. 385-390.

Author index

Subject index

This index is compiled from the keywords assigned to the papers, edited and extended as appropriate. The page references are to the first page of the relevant paper.

Printed and bound by CPI Group (UK) Ltd, Croydon, CR0 4YY

01/11/2024

01782621-0010